SCOURING

INTERNATIONAL ASSOCIATION FOR HYDRAULIC RESEARCH
ASSOCIATION INTERNATIONALE DE RECHERCHES HYDRAULIQUES

2

HYDRAULIC STRUCTURES
DESIGN MANUAL

HYDRAULIC DESIGN CONSIDERATIONS

SCOURING

H.N.C. BREUSERS
Delft Hydraulics, Delft, Netherlands

A.J. RAUDKIVI
Howick, New Zealand

A.A. BALKEMA / ROTTERDAM / BROOKFIELD / 1991

Photo on the binding: Collapsed Bulls Bridge site looking upstream. See Figure 5.16, p. 82.

Authorization to photocopy items for internal or personal use, or the internal or personal use of specific clients, is granted by A.A.Balkema, Rotterdam, provided that the base fee of US$1.00 per copy, plus US$0.10 per page is paid directly to Copyright Clearance Center, 27 Congress Street, Salem, MA 01970, USA. For those organizations that have been granted a photocopy license by CCC, a separate system of payment has been arranged. The fee code for users of the Transactional Reporting Service is: 90 6191 983 5/91 US$1.00 + US$0.10.

Published by
A.A.Balkema, P.O.Box 1675, 3000 BR Rotterdam, Netherlands
A.A.Balkema Publishers, Old Post Road, Brookfield, VT 05036, USA

ISBN 90 6191 983 5

© 1991 A.A.Balkema, Rotterdam
Printed in the Netherlands

Contents

List of symbols		VII
1	Introduction *H.N.C. Breusers & A.J. Raudkivi*	1
2	Basic concepts of soil erosion and sediment transport *A.J. Raudkivi*	7
3	Scour in rivers and river constrictions *H.N.C. Breusers*	37
4	Scour around spur dikes and abutments *H.N.C. Breusers*	51
5	Scour at bridge piers *A.J. Raudkivi*	61
6	Scour by jets, at high-head structures and at culvert outlets *H.N.C. Breusers*	99
7	Scour below low head structures *H.N.C. Breusers*	123
Subject index		143

List of symbols

Symbol	Description	Unit
A	area of cross-section	m²
B	width of river	m
B_s	width of scour	m
B_u	half width of jet	m
b	width of structure or pier	m
C	Chézy roughness coefficient	m$^{1/2}$/s
c	concentration (usually by volume)	—
D_u	diameter of jet	m
D	diffusion coefficient	m²/s
d	grain size $\bar{d}$ = mean grain size	m
D_*	dimensionless grainsize $D_* = d(\Delta g/v^2)^{1/3}$	—
d_{50}	median grain size	m
Fr	Froude number = $U/\sqrt{gy_o}$	—
f	Darcy-Weissbach friction factor	—
f	Lacey silt factor	—
g	acceleration due to gravity	m/s²
H	total energy head, head loss	m
h	water depth, height (e.g. dune height)	m
k	equivalent roughness height (Nikuradse)	m
l	length of pier	m
n	porosity, Manning coefficient	—
P	pressure	N/m²
Q	discharge of water	m³/s
Q_s	discharge of sediment (volume of grains)	m³/s
q	discharge of water/m width	m²/s
q_B, q_S, q_T	discharge of sediment/m width (B = bed load, S = suspended load, T = total load)	m²/s
R	hydraulic mean radius	m
Re	Reynolds number	—
r	radius	m

VIII List of symbols

S	hydraulic gradient	—
s	scale factor	—
S.F.	shape factor	—
U	depth-averaged flow velocity	m/s
U_c	average flow velocity at initiation of sediment motion	m/s
U_o	exit velocity of jet	m/s
V	average flow velocity in cross section	m/s
u_*	shear velocity	m/s
u_{*c}	shear velocity at initiation of sediment motion	m/s
u	mean velocity of at elevation y	m/s
w	fall velocity of sediment	m/s
y_o	uniform flow depth	m
y_r	regime depth	m
y_s	scour relative to river bed	m
y_{se}	equilibrium value of y_s	m
z	bed level	m
α	angle of attack	—
Δ	$(\rho_s - \rho)/\rho$	—
ε_s	turbulent diffusion coefficient for sediment	m^2/s
κ	Von Kármán constant	—
λ	wave length, meander length, dune length	m
μ	dynamic viscosity	Ns/m^2
ν	kinematic viscosity	m^2/s
ρ, ρ_s	density of water, sediment	kg/m^3
σ_g	geometric standard deviation of the grain size distribution	—
τ_o	bed shear stress	N/m^2
τ_c	bed shear stress at initiation of sediment motion	N/m^2
θ	dimensionless shear stress	—

CHAPTER 1

Introduction

H.N.C. BREUSERS & A.J. RAUDKIVI

1.1. GENERAL

The aim of this monograph is to assemble and interpret information from both laboratory and field which would help the designer as he seeks to produce safe and economic structures in streams with erodible beds. Such structures often induce serious erosion problems because through interference with the flow pattern they lead to increased erosion.

Scouring is a natural phenomenon caused by the flow of water in rivers and streams. It is most pronounced in alluvial materials, but deeply weathered rock can also be vulnerable in certain circumstances. Experience has shown all too often that scouring can progressively undermine the foundation of a structure. Because complete protection against scouring is usually prohibitively expensive, the designer must seek ways to guide and control the process so as to **minimize** the risk of failure. Guidance comes both from controlled studies in laboratories and from field experience, both the successes and particularly the failures. Despite much study, the principles of analysis of scouring are not well established, and the results of various investigations often show differing trends. Hence, the following chapters can only offer guidance to a complex subject rather than prescriptions of established procedure.

Scouring occurs naturally as part of the morphologic changes of rivers and as the result of man-made structures. The development of river valleys reveals such activity through millenia, long before man's efforts had any appreciable impact on them. In recent times, nonetheless, the addition of many types of structures has greatly altered river regimes, and significant impacts on the transport and deposition of sediment have resulted. Most structures increase these processes, at least locally and often to the detriment of the river regime. The designer must therefore seek to understand the scouring process and to study the consequences of a given structure and its effect on the larger processes of river morphology.

Major floods and the scour they produce can be so large as to present daunting challenges to the engineer. Natural scouring can cause dramatic changes in the plan, cross section and even location of a river. Bridges are sometimes left behind "in the dry" as a consequence of natural channel changes. Bends and narrows of a river channel tend to scour during floods and fill at low flows. Rivers with sand beds display complex systems of bed features which translate downstream with appreciable variation in local bed levels. Bank erosion can occur as a part of the slow translation of the meander pattern and can lead to natural cut-off of bends with an associated degradation of the bed upstream. In wide braided river channels, where low flows occupy only a few small channels, substantial shifts of the deep water channel can occur during floods. These are but a few of many erosional/depositional features that determine river-channel morphology. Thus the designer can be confronted with two kinds of problems. Firstly, he may need to stabilize the location of a channel for his design to function correctly, in which case he must combat the river's natural tendency to wander. Secondly, because structural modifications usually increase the tendency to scour, at least locally, he has to protect the structure against the danger of being undermined.

Designers must operate without a comprehensive theory like those for the physical structures that are involved. They are unable to predict with confidence the depths and locations of scour caused by spillways, weirs, bridge piers, abutments, culverts and other such devices. Two phenomena intimately associated with the scouring are so complex as to preclude efforts at full understanding and hence the achievement of accurate solutions:

— turbulence, with its great complexity and variability, and
— sediment transport with its strong dependence on the complex interactions with turbulent flows.

Structural failures induced by scour usually occur in extreme cases of unsteady flows interacting with a given structure and with changing channel conditions. An additional complexity is the composition of the sediment which is usually a mixture of alluvial sands, clay and weathered rock. A stretch of river can display alluvium, clay banks, rock outcrops and bars of sand and gravel. The designer provides a further complexity by producing a wide variety of structural shapes and alignments. The most dramatic and perhaps the most useful of the various types of guidance available are the failures that have occurred as the result of inadequate protection against scour. The engineer is required as best he can to relate the experiences and concepts of others to the situation which confront him. The various chapters which follow provide guidance in this difficult process.

1.2. DEFINITIONS

The scour which may occur at a structure can be divided into the following categories:

A. *Types of scour*

General scour occurs in a river or stream as the result of natural processes irrespective of whether a structure is there.

Constriction scour occurs if a structure causes the narrowing of a water course or the rechanneling of berm or flood plain flow.

Local scour results directly from the impact of the structure on the flow. This scour, which is a function of the type of structure, is superimposed on the general and constriction scour.

B. *Scour in different conditions of transport*

Clear-water scour occurs if the bed material in the natural flow upstream of the scour area is at rest. The shear stresses on the bed some distance away from the structure are thus not greater than the critical or threshold shear stress for the initiation of particle movement.

Live-bed scour, also referred to as scour with bed material sediment transport, occurs when the flow induces a general movement of the bed material. That is, the shear stresses on the bed are generally greater than the critical one. Equilibrium scour depths are reached when the amount of material removed from the scour hole by the flow equals the amount of material supplied to the scour hole from upstream.

1.3. MODELLING TECHNIQUES

Many projects, especially the larger and more complex ones, require model studies. Techniques available at present involve numerical and physical models. The former will have limited predictive potential although their utility will increase with the formulation of adequate theoretical structures to support them. The latter, the small-scale physical model, is at present a much more powerful tool.

Techniques for assuring acceptable similarity between model and prototype have evolved, and are based on several criteria:

(1) Geometrical conformity (no distortion) because significant aspects of the flow are three-dimensional.
(2) Froude number similarity: velocity scale (s_U) varying as the square root of the length scale $(s_L)^{1/2}$ because free surface phenomena are usually important.
(3) Similarity of sediment transport, characterized by:

$$s_U = s_{U_c}$$

in which U_c is the velocity at initiation of sediment motion.

The second criterion can be relaxed for flow around bridge piers at low Froude numbers, in that some exaggeration of the velocities in the model is possible. Kolkman (1982) describes an interesting technique to allow deviation from the Froude number scaling to fulfill the third criterion. The free surface is measured at the proper Froude scale, and then an artificial ceiling is shaped accordingly. Scouring tests can then be done at exaggerated velocities. Experiments showed that the velocity could be increased up to a factor of two. Thus geometric similarity is maintained together with greater scour potential in the model.

The third criterion is not difficult to fulfill for coarse non-cohesive material as U_c varies with $d^{1/2}$ for $d > 2$ mm. Thus sediment size can be scaled at the length scale as long as $d > 2$ mm in the model. With beds of finer sediments in nature, problems arise in modelling. Within certain limits, fine sand can be simulated with light-weight materials like bakelite, lignite, polystyrene etc. Studies on time-dependent scour in fine sediments (see Chapter 7) have shown that criterion 3 can be relaxed if similar values of scour depth to original water depth are compared and the corresponding time scale is taken into account.

Reproduction of rocks for scouring studies is difficult. Fissured rock can be simulated by concrete cubes. Cube size is scaled directly with the length scale from the size of the fissured blocks. For non-fractured rock, cohesive materials can be employed in the model having a critical velocity which corresponds to the estimated critical velocity in nature at velocity scale (Johnson 1977, Gerodetti 1979). More general discussions of modelling techniques have been presented by Yalin (1971), Kobus (1980) and Novak and Čábelka (1981).

1.4. OUTLINE OF THE MONOGRAPH

Scouring is due to gradients in sediment transport caused by the presence of structures or disturbances in the flow field. Knowledge of sediment transport processes is therefore essential. In Chapter 2 those aspects of sediment transport which are important for the scouring process are discussed in an introductory way.

The scouring process greatly depends on the geometry of the structure under consideration. General scour due to changes in river morphology or geometry is discussed in Chapter 3. In Chapter 4 data on scour around spur dikes and abutments is reviewed and recommendations for design are given. Chapter 5 follows with scour at bridge piers, where knowledge is relatively abundant and good guidelines for design can be given. Scour near structures is discussed in Chapters 6 and 7 for structures producing jet-type flows and low-head structures, respectively.

Where possible, guidelines for design are given. It must be realized, however, that most knowledge is based on small-scale research and that extrapolation to prototype conditions must be done with care.

ACKNOWLEDGEMENT

The authors wish to thank in particular Dr. John McNown, who has carefully reviewed all material and has greatly improved the clarity of the text.

REFERENCES

Gerodetti, M. 1979. *Auskolkung eines felsigen Flussbettes. Modellversuche mit bindigen Materialen zur Simulation des Felsen.* E.T.H. Zürich, VAWAG, Arbeitsheft 5.

Johnson, G.M. 1977. Use of a weakly cohesive material for scale model scour studies in flood spillways design. *Proc. 17th IAHR Congress, Baden-Baden* 4; 509.

Kobus, H. (ed.) 1980. *Hydraulic modelling.* German Association for Water Resources and Land Improvement, Bull. 7 (in co-operation with IAHR, Verlag Paul Parey and Pitman).

Kolkman, P.A. 1982. An artificial ceiling for free surface flow reproduction in scale modelling of local scour. *Proc. Int. Conf. on Hydraulic modelling of Civil Engineering Structures, Coventry;* 397–410.

Novak, P. & S. Čábelka 1981. *Models in hydraulic engineering: Physical principles and design applications.* Pitman, Boston.

Yalin, M.S. 1971. *Theory of hydraulic models.* MacMillan, New York.

CHAPTER 2

Basic concepts of soil erosion and sediment transport

A.J. RAUDKIVI

Sediment transport, two-phase flow and loose boundary hydraulics are some of the names used to identify problems of interaction between a fluid flow and a granular material mixed into the flow and forming some of the boundaries of the flow which usually change with flow conditions. The major exception is pipeline conveyance where the boundaries are usually formed by the pipe. These transport problems occur in a multitude of natural and industrial processes, from soil erosion, materials handling, industrial processes, to burning of pulverized coals in furnaces. The crucial feature is the interaction between fluid and the sediments, that is, the sediment problems cannot be treated in isolation from hydro- or aerodynamics. These are subjects in their own right and have many unsolved problems, in particular in areas of turbulence, boundary layer, wave motion and diffusion, all of which may be associated with the problem of sediment transport.

The mechanics of interaction of fluid flow and sediment is subject of numerous papers and a number of textbooks, e.g. Bogardi (1974), Garde and Ranga Raju (1977, 1985), Graf (1971), Raudkivi (1967, 1976, 1982, 1990), Simons and Sentürk (1976), Yalin (1972). The results of these interactions in nature form the branch of science known as river morphology, Leopold et al. (1964), Richards (1982).

2.1. SEDIMENT CHARACTERISTICS

For convenience, the sediments forming the boundaries of a flow are subdivided into cohesive and non-cohesive sediments, although there is a fairly broad transition range. In alluvial or non-cohesive sediments the particle or grain size and weight are the dominant parameters for sediment movement and transport. Non-cohesive sediments have a granular structure and do not form a coherent mass. However, the properties of alluvial soils change drastically with increasing clay content (fraction of soil composed of particles smaller than 2 μm). In most soils, clay assumes control of soil properties already at 10 per cent clay content. In cohesive soils the electro-chemical interactions dominate and the size and weight of an individual particle may be of little importance. Cohesive soils form a coherent mass. For erosion and transport of cohesive soils reference is made to Raudkivi (1982, 1984, 1990).

Soils are classified according to particle size, for example British Standard BS1377: 1975, as follows:

Very fine clay	0.24–0.5	μm	Very fine gravel	2–4	mm
Fine clay	0.5–1.0	μm	Fine gravel	4–8	mm
Medium clay	1–2	μm	Medium gravel	8–16	mm
Coarse clay	2–4	μm	Coarse gravel	16–32	mm
Very fine silt	4–8	μm	Very coarse gravel	32–64	mm
Fine silt	8–16	μm	Small cobbles	64–128	mm
Medium silt	16–31	μm	Large cobbles	128–256	mm
Coarse silt	31–62	μm	Small boulders	256–512	mm
Very fine sand	62–125	μm	Medium boulders	512–1024	mm
Fine sand	125–250	μm	Large boulders	1024–2048	mm
Medium sand	250–500	μm	Very large boulders	2048–4096	mm
Coarse sand	0.5–1.0	mm			
Very coarse sand	1–2	mm			

A very similar classification was proposed by the Subcommittee on Sediment Terminology of the American Geophysical Union (Vanoni, 1975, page 20, Table 2.1). The common definitions for particle size are:

sieve diameter – opening size of mesh (may include oblong particles).
sedimentation diameter – diameter of a sphere of the same density and the same fall velocity in the same fluid at the same temperature as the given particle.
nominal diameter – diameter of a sphere of equal volume.
triaxial dimensions a, b and **c**, where c is the shortest of the three mutually perpendicular axes of the particle.

The sediment normally consists of a distribution of particle sizes. Figure 2.1 shows the two common presentations of the particle size distribution, the cumulative distribution curve as the log-normal and the log-probability plot. The 50% diameter on the log-probability plot is called the geometric mean diameter d_g and the geometric standard deviation is defined as

$$\sigma_g = (d_{84.1}/d_{15.9})^{1/2} \quad \text{or} \quad \sigma_g = \frac{d_{84.1}}{d_{50}} = \frac{d_{50}}{d_{15.9}} \tag{2.1}$$

Many natural deposits plot as one or more straight line segments on log-probability paper. A two segment distribution can be shown to arise from superposition of two particle size distributions.

The mean diameter, $\bar{d}$, is given by arithmetic mean of the size distribution. The values of $\bar{d}$ and d_g are related by

$$\bar{d} = d_g \exp[0.5 \ln^2 \sigma_g] \tag{2.2}$$

Several definitions of an "effective" grain size, d_e, have been proposed, e.g., $d_e =$

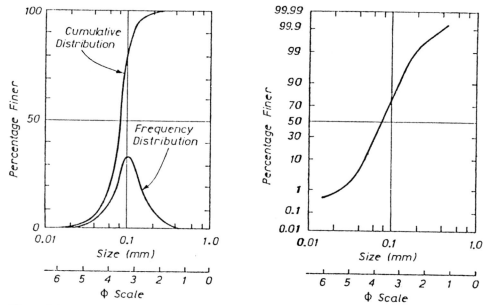

Figure 2.1. Ways of presenting grain size distributions.

$(d_{10}+d_{20}+\cdots+d_{90})/9$, or from equal areas on the plot of $1/d$ as the ordinate and percentage passing as the abscissa, i.e. the horizontal line at $1/d_e$ divides the area under the grading curve into equal parts. In the same sense a mean grain size has been proposed

$$\bar{d} = \frac{\sum_{p=0}^{100\%} \bar{d}_i p_i}{\sum_{p=0}^{100\%} p_i}; \quad \bar{d}_i = \tfrac{1}{2}(d_i + d_{i+1}) \tag{2.3}$$

where p_i is the percentage weight of size d_i.

Further characteristics of sediment grading in use, particularly in river morphology, are the sorting coefficient, sorting index and uniformity index.

Generally, the grain-size distributions near the source, like debris slides, are strongly skewed towards the larger fractions (percentage by weight versus size). During the transport processes the distributions gradually transform and reach a normal (gaussian) form after a sufficient distance, i.e., the cumulative distribution plots as a straight line on log-normal paper.

The phi-index scale is often used by sedimentologists and is defined by

$$\phi = -\log_2 d_{(mm)} = -\log d / \log 2 \tag{2.4}$$

For example, for $d = 0.25$ mm, $\phi = 2$.

The size of the sample for analysis depends on the size of the large particles present. The ASTM recommends a sample size M in kg

$$M = 0.082 b^{1.5} \tag{2.5}$$

where b is the maximum intermediate triaxial dimension in mm.

There are two simple criteria by which it may be assessed if a sediment mixture can be treated as though it were uniform. If $d_{95}/d_5 < 4$ or 5 the sediment is uniform from the hydraulic point of view, and similarly if $\sigma_g < 1.35$ the sediment may be considered uniform. For mixtures which do not meet these criteria, the non-uniformity of grain size reduces the resistance to flow and transport of sediment. The coarser grains tend to armour the surface which leads to reduction of effective roughness. Bed forms too in non-uniform sediment are lower and flatter due to armouring effects and present a reduced roughness to flow compared to that over a bed of uniform grains. Due to armouring the sediment transport can vary with time under constant flow rate and even go to zero. The erosion of non-uniform sediment is a more complex problem than that of sediments of uniform grain size. The erosion process depends on the spatial distribution of particle size in the surface layers. Locally this influences the formation of armoured beds (discussed below) while on the larger scale it can affect the whole river regime. For this, reference may be made to Leopold et al. (1964), Allen (1965) and Reineck and Singh (1973) who discuss the variation in fluvial sediments in terms of the channel maturity and sinuosity. In general, the mean particle size of a channel bed decreases and the uniformity of grain size increases as the channel slope decreases. The least sorted sediments are in the fluvial deposits (fans) at the foot to the mountains. These sediments have not yet been sorted by preferential transport over larger distances.

For hydraulic calculations with beds of non-uniform grains one needs a measure of an effective grain size. This has been discussed, for example, by Christensen (1969) and Irvine and Sutherland (1973). However, the effective particle size problem is still unresolved and the user must select his own definition. The use of d_{50} is adequate for relatively narrow grain size distributions. As the grading broadens, and particularly for roughness calculations, a value of d_{75} or d_{80} is thought to be more appropriate.

A further important parameter is the *particle shape*. One of the definitions in use is the shape factor

$$SF = c/(ab)^{1/2} \tag{2.6}$$

where c is the smallest of three triaxial dimensions. The *fall velocity* features in all the descriptions of suspension and sedimentation. For a spherical particle the fall velocity is

$$w = \left[\frac{4}{3} \frac{1}{C_D} g d \Delta \right]^{1/2} \tag{2.7}$$

where $\Delta = (\rho_s - \rho)/\rho$ and ρ_s is the density of the grain and C_D is the drag coefficient which is a function of the Reynolds number $Re = wd/v$ where v is the kinematic viscosity. The fall velocity depends on the shape of the particle and the Reynolds number.

The fall velocity of naturally worn quartz grains in terms of nominal diameter (which is readily determined from weight and density), according to US Inter-Agency Committee (1957) is shown in Figure 2.2. For naturally worn sediments over the range of about 0.2 to 20 mm the sieve diameter is approximately 0.9 times the nominal diameter.

Average values for quartz sands in water at 20°C appear to fit reasonably well to

$$w_{(mm/s)} \approx 663 d^2_{(mm)}; \quad d < 0.15 \text{ mm}$$

$$w_{(mm/s)} \approx 134.5 d^{0.52} \approx 134.5 \sqrt{d_{(mm)}}; \quad d > 1.5 \text{ mm}$$

with a transition region between $0.15 \leq d \leq 1.5$ mm

d mm	0.15	0.2	0.3	0.4	0.5	0.6	0.7	0.8	0.9	1.0	1.2	1.5
w mm/s	14.8	21.1	36.1	50.0	64.0	76.4	88.6	99.0	110.0	121.0	137.3	166.0

where d is the nominal diameter.

The effective fall velocity of a sediment mixture can be expressed as

$$w = \frac{\Sigma p_i w_i}{\Sigma p_i} \tag{2.8}$$

where p_i and w_i are the weight and fall velocity of the grains in the size range i.

The influence of particle concentration on the fall velocity can be expressed as

$$w = w_o(1-c)^\beta \tag{2.9}$$

where c is volumetric concentration, w_o is the fall velocity of a single grain and β is a function of $Re = wd/v$ and particle shape. It can also be expressed as a function of the nondimensional grain size

$$D_* = (\Delta g d^3/v^2)^{1/3} \tag{2.10}$$

where $\Delta = (\rho_s - \rho)/\rho$. For common natural grains (SF ≈ 0.7), the value of $\beta = 4.65$ for $D_* < 40$ and $\beta = 2.35$ for $D_* > \approx 800$, with a transition defined by $\beta = 7.48 D_*^{-0.129}$.

The volumetric and weight concentration c and c_w, are related through

$$c = \frac{c_w/S_s}{(1-c_w)+c_w/S_s} \tag{2.11}$$

where $S_s = \rho_s/\rho$.

Sediment concentration also affects the effective viscosity of water. An approximate expression is

$$\frac{\mu_{susp.}}{\mu} = 1 + k_1 c + k_2^2 c^2 + k_3^3 c^3 + \tag{2.12}$$

where $k_1 \approx 2.5$ for $c < 2$–3%. As a first approximation $k_1 = k_2 = k_3$.

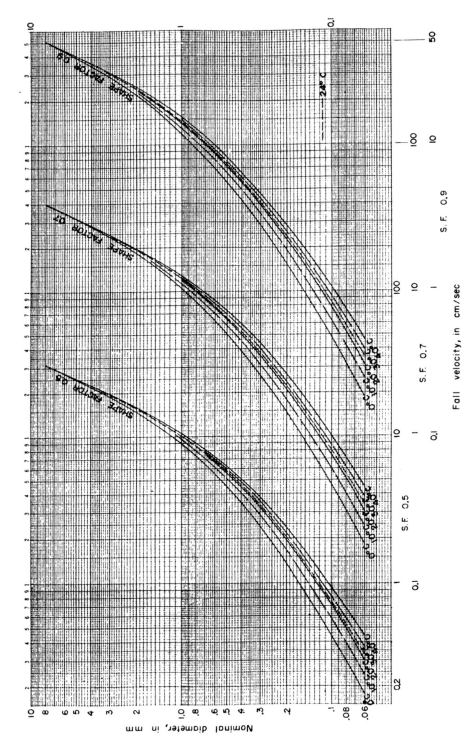

Figure 2.2. Relation of nominal diameter and fall velocity for naturally worn quartz particles with shape factors (SF) of 0.5, 0.7, and 0.9 according to U.S. Inter-Agency Committee on Water Resources, Subcommittee on Sedimentation, 1957.

2.2. SEDIMENT ENTRAINMENT

The initiation or threshold of movement of a particle due to the action of fluid flow is defined as the instant when the applied forces due to fluid drag and lift, causing the particle to move, exceed the stabilizing force due to gravity. For uniform sediments in unidirectional flow this condition is best defined by the Shields curve, Figure 2.3, which defines the threshold in terms of the entrainment function

$$\theta_c = \frac{\tau_c}{\rho g \Delta d} = \frac{u_{*c}^2}{g \Delta d} \qquad (2.13)$$

which is a dimensionless shear stress, and the grain Reynolds number $Re_* = u_* d/\nu$ where $u_* = (\tau_o/\rho)^{1/2}$ is the shear velocity, and τ_o and τ_c are the bed shear stress and its critical value respectively, i.e. $\tau_o = (\rho g y_o S)$ where y_o is the depth of flow and S is the slope.

A drawback of the Shields diagram is that the shear velocity appears on both axes. In terms of the dimensionless particle size $D_* = (\Delta g d^3/\nu^2)^{1/3}$ the data on threshold is described by

$$D_* = 2.15\, Re_*; \quad Re_* < 1 \qquad (2.14a)$$
$$D_* = 2.5\, Re_*^{4/5}; \quad 1 > Re_* < 10 \qquad (2.14b)$$
$$D_* = 3.8\, Re_*^{5/8}; \quad Re_* > 10 \qquad (2.14c)$$

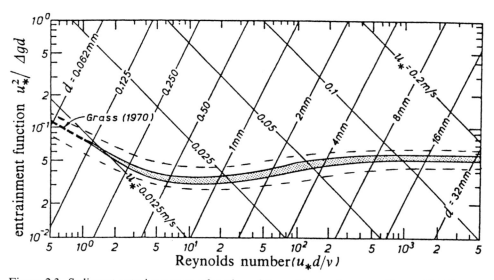

Figure 2.3. Sediment entrainment as a function of Reynolds number according to Shields. The shaded band indicates the spread of data by Shields, the dashed lines the envelope to most of the published data.

The Shields curve applies to uniform sediment at average levels of turbulence. Most alluvial sediments, with the exception of fine sands, have a broad grain size distribution. The geometric standard deviation of river gravels is normally in the order of 4. The erosion of such beds under certain conditions can lead to the formation of an armour layer on the bed surface, also known as paving, which protects the bed.

The armouring is a function of the applied bed shear stress. Figure 2.4 illustrates the gradings of the armour layer at different applied shear stresses. As the shear stress is increased, a condition will be reached where the surface does not armour any more and all particles are indiscriminately transported by the flow. Up to this limiting shear stress, or critical shear stress of the armour layer, the d_{50} particle size of the armour layer, d_{50a}, increases with shear stress. The development as illustrated by Figure 2.4 is insensitive to the original grading. The process is controlled by the coarse fraction and the d_{max} size. The grading affects the time scale of development. Initially d_{50a} increases approximately in proportion to shear stress but becomes insensitive to shear stress near the limiting shear. Chin (1985) found that the ratio of d_{max}/d_{50a} approached a lower limit

$$\frac{d_{max}}{d_{50a}} \approx 1.8 \tag{2.15}$$

This limit is approached asymptotically and is reached when $u_*/u_{*ac} \geqq 0.9$ where u_{*ac}

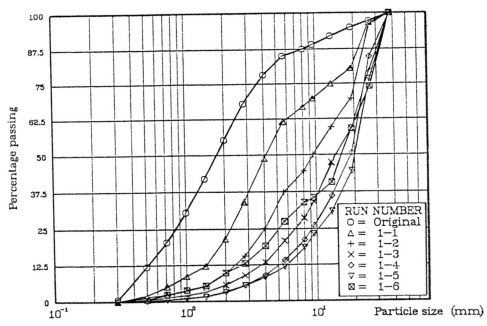

Figure 2.4. Variation of equilibrium armour layer grain size distribution with increasing shear velocity. The shear velocities were 41, 53, 66, 76, 85 and 95 mm/s for runs 1 to 6, respectively, with d_{50a} of 4.2, 9.5, 14.5, 19.0, 20.5 and 16.0 mm.

is the limiting critical armour layer shear velocity. Equation 2.15 implies $d_{84}/d_{50} \approx 1.5$ or approximately uniform grading of the coarse fractions. The critical dimensionless shear stress θ_{ca} is a function of $(d_{50a})_{max}/d_{50}$, and can be fitted (following White and Day, 1982) by

$$\frac{\theta_{ca}}{\theta_c} = \left\{ 0.4 \left(\frac{(d_{50a})_{max}}{d_{50}} \right)^{-0.5} + 0.6 \right\}^2 \tag{2.16}$$

where the original bed material is characterized by d_{50} and $\theta_c \approx 0.05$.

The characteristic size d_{max} of the bed material is difficult to estimate and varies from sample to sample. The armour layer is clearly not affected by an occasional extra-large stone.

One method to obtain d_{max} is to extrapolate the grading curve on the basis of the last two or three data points to the 100% passing size. This approach relies also to some extent on judgement. The d_{max}-value could also be estimated by using two sieves, such that all particles just pass the coarser sieve. If the sieves are in the $2^{1/4}$-series the error involved by assuming d_{max} to be equal to the coarser sieve would be less than 20% i.e. $(2^{1/4}-1)$. If an observed grading of the limiting armour layer then extends to the same d_{max} size (see Figure 2.4), this would be a confirmation for the d_{max} estimate.

The major effect of the armouring is that, at all shear stresses less than the critical armour layer shear stress, the sediment transport rate initially decreases with time as the surface armouring develops. Usually these armour clusters here and there fail, giving rise to locally increased erosion until the armour there builds anew, and this maintains on average a steady bed load transport rate. However, when there is a decrease in flow rate the sediment transport rate can effectively go to zero. Indiscriminate steady bed load transport occurs only if the bed shear stress exceeds the limiting armour layer shear stress.

The threshold of movement of a grain is also strongly influenced by its position in the surface, Figure 2.5. A grain from a co-planar layer is obviously more difficult to move than one lying on top of the others. In particular, it illustrates the importance of the underlying bed for the stability of individual elements, e.g., rock armour units. An exposure of only half a diameter above the general bed level reduces the threshold for movement by about six times compared to that in a coplanar arrangement. An exposed particle can be transported over the bed without disturbing it, by a bed shear stress, τ_o, equal or less than the critical shear stress, τ_c, of the bed material in general. This transport can be of coarser grains as well as finer ones. For example, the deposits left after floods on floodways of steeper rivers are often dotted with large stones on the surface of the deposit. The preferential transport can be of coarser as well as finer grains than the bed material in general. If the shear stress, τ_i, required to move a particular grain or stone is greater than $\tau_o > \tau_c$ the bed material moves but the exposed stone is too large to be moved. It acts as an obstruction to flow, a scour hole develops in front of it and the stone becomes embedded in time. The most important case is when $\tau_c > \tau_o > \tau_i$, then the exposed particles will move over the bed which itself remains undisturbed. Larger particles overpass a bed composed of smaller particles by rolling and sliding

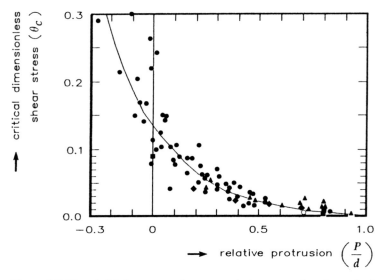

Figure 2.5. Threshold shear stress θ_c versus relative protrusion above the bed. ● - Data by Fenton and Abbott (1977); ▲, ■, ♦, ○, - Data by Chin, University of Auckland.

whereas smaller overpassing particles show a saltating or bouncing form of movement. These are the conditions for sediment sorting and preferential transport.

The above features of overpassing or embedding are made very much more complex by the enormous number of possible grain shapes. The simple picture is also modified by areal groupings of larger stones which can give the bed material, and each other, substantial protection. However, the fines could still be entrained from between the stones by local wakes and vortices, so-called winnowing.

Numerous formulae exist for the "critical velocity" for particle movement. The critical velocity can also be calculated for given conditions from the critical shear stress or shear velocity. Thus, the mean critical velocity is

$$U_c = u_{*c} C/\sqrt{g} \quad \text{or} \quad U_c = u_{*c}\left(5.75 \log \frac{y_o}{2d} + 6\right) \tag{2.17}$$

or at a given elevation

$$u_c = 5.75 u_{*c} \log \frac{y}{y'} \tag{2.18}$$

where C is the Chézy coefficient and y' is the elevation at which the logarithmic velocity distribution has zero velocity. The problem with the critical mean velocity is that the bed shear stress for the same mean velocity decreases with increasing depth of flow. The use of a critical bed velocity u_b suffers from the difficulty of definition of the elevation where it is to be measured and its relationship to the mean velocity. Probably the best known of the early contributions on critical velocity is the Hjulström (1935) curve. Neill (1967, 1968) specified a "conservative design curve" for uniform coarse material (gravel) in terms of critical mean velocity U_c as

$$\frac{U_c^2}{\Delta gd} = 2.0 \left(\frac{d}{y_o}\right)^{-1/3} \tag{2.19}$$

where y_o is the depth of uniform flow.

Many soils are *cohesive* and most fine-grained sediments, such as fine silts, possess some cohesive properties. Yet, the understanding of the physics of erosion of cohesive soils is very limited and no rational models exist which are capable of quantitative predictions of erosion rates of cohesive soils in general. The clay content has a controlling effect on soil properties, only about 10% of clay will assume complete control.

The erosion products of cohesive soils are transported mainly as suspended and wash load and have a significant effect on the water quality downstream.

Space does not permit here a more detailed discussion of the erosion of cohesive soils. For an introduction, the reader is referred to Raudkivi (1982, 1984, 1990). Fine grained deposits under water or intertidal areas can also be rendered "cohesive" by biological effects, for example, by growth of algae.

2.3. CHANNEL ROUGHNESS

The alluvial channel differs from a fixed boundary open channel through its ability to alter the channel geometry and roughness to flow. In a straight fixed boundary channel the resistance to flow is related to the boundary roughness which is independent of flow. The roughness of an alluvial channel varies with flow and is only indirectly related to the grain size and grain size distribution of the boundary material. No method exists to date by which one could reliably calculate the roughness and energy loss in an alluvial channel. The multitude of empirical methods illustrates that the solution has not yet been found.

For a given slope, which is largely controlled by the slope of the land, the resistance to flow manifests itself in flow depth. The latter is an important design parameter for all hydraulic structures, with or without scour. For a two-dimensional turbulent flow over a fixed hydraulically rough surface the mean velocity at elevation y above the bed is

$$\frac{u}{u_*} = 5.75 \log\left(\frac{30.2y}{k}\right) = 5.75 \log\frac{y}{k} + 8.5 \tag{2.20}$$

which equals the mean velocity, U, at $y = 0.368 y_o \approx 0.4 y_o$, where k is the equivalent sand roughness height of the surface and y_o is flow depth. Comparison with the Chézy formula

$$U = C\sqrt{RS} \tag{2.21}$$

where R is the hydraulic mean radius and S is the hydraulic gradient, shows that

$$C = 18 \log\left(\frac{y_o}{k}\right) + 18.8 = 18 \log\left(\frac{11.1R}{k}\right) \qquad (2.22)$$

Plotted on log-log scale this relationship for C has a slope of $\tan^{-1}(1/6)$ at about $y_o/k = 30$, i.e., at this relative roughness

$$C = K\left(\frac{y_o}{k}\right)^{1/6} \qquad (2.23)$$

Substituting into the Chézy formula yields

$$U = \frac{K}{k^{1/6}} y_o^{2/3} S^{1/2} \qquad (2.24)$$

which yields the Manning or Strickler formula

$$U = \frac{1}{n} R^{2/3} S^{1/2} \qquad (2.25)$$

where n is the Manning coefficient.

The Chézy and Manning coefficients and the Darcy-Weisbach friction factor f are related by

$$C = \frac{1}{n} y_o^{1/6}; \quad \frac{C}{\sqrt{g}} = \left(\frac{8}{f}\right)^{1/2} \qquad (2.26)$$

where $\tau_o = (f/4)\rho U^2/2$. \hfill (2.27)

Strickler (1923) expressed n for natural channels as

$$n = \frac{1}{21.1} d^{1/6} = 0.0474 d^{1/6} \qquad (2.28)$$

where d is the grain size. For this Lane proposed the d_{75}-size. If

$$n = d_{(m)}^{1/6}/24 = 0.042 d_{(m)}^{1/6} = 0.013 d_{(mm)}^{1/6} \qquad (2.29)$$

the Manning formula yields

$$\frac{U}{u_*} = 7.66 \left(\frac{R}{d}\right)^{1/6} \qquad (2.30)$$

giving results almost identical to those from the logarithmic formula

$$\frac{U}{u_*} = 5.75 \log\left(\frac{y_o}{k}\right) + 6 \qquad (2.31)$$

The Manning formula is consistent with the logarithmic velocity distribution in the neighbourhood of $y_o/k \approx 30$. In the region of $y_o/k = 10$ the Chézy coefficient $C \sim (y_o/k)^{1/4}$ yielding $U \sim R^{3/4} S^{1/2}$ which is the form proposed by Lacey (1929).

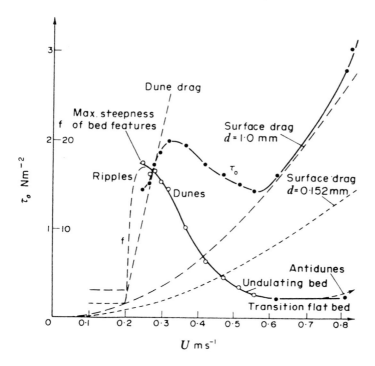

Figure 2.6. Variations of bed shear stress τ_o and Darcy-Weisbach friction factor f with mean velocity U in flow over a fine sand bed.

In an alluvial channel the roughness depends on

(a) the nature of bed material, its grading and properties (particularly shape) and on the spatial variation of these in the channel, and
(b) the flow depth and velocity (or shear velocity) which determine the nature of the bed features for a given bed material.

The inherent difficulty with the estimation of the steady state flow depth in an alluvial channel arises from the variation of boundary roughness or shear stress with the velocity of flow, as illustrated for a sand bed in a laboratory flume in Figure 2.6.

The changes in roughness due to bed features can be substantial in flow over fine sand beds as illustrated by laboratory data. Darcy-Weisbach f-value can increase about tenfold from the initial flat bed value and it returns to almost the same value at the transition flat bed conditions. The combined grain roughness and that caused by the sediment in motion is seen to equate to approximately $6d_{50}$. This may well be a more representative value for the transition flat bed in general than the idealized flat bed of grain roughness only. The initial flat bed may be rare in nature but the transition flat bed is a common occurrence. The first observation is that for a given f-value there may be three values for flow velocity. In nature the slope of the stream cannot change significantly with flow rates. Therefore, the effect of the changes in roughness is a substantial variation in flow depth. Since the bed shear stress for approximately constant channel slope is proportional to flow depth, the stage-discharge curves also vary with the changing bed features; they show a "discontinuity" where the flow passes through the transition flat bed conditions, from the *lower regime* to the *upper regime*. Examples can be found in papers by Simons et al. (1962), Nordin (1964) and others.

The form drag over well-formed sand dunes and ripples is high because their steep lee faces (about 30°) approximate to a negative step with a substantial lee eddy and expansion loss. In nature the range of f-values is likely to be smaller, because the bed features are more random in size and distribution, but the form drag over sand beds is still a significant component of energy loss.

Whenever the velocity over an alluvial bed exceeds the threshold value, bed features develop. Conventionally these are subdivided into ripples, dunes and antidunes. There are numerous presentations which attempt to delineate conditions leading to a particular form of bed features, for example, Figure 2.7 by Simons et al. (1964).

Ripples develop at small values of shear stress excess ($\tau_o - \tau_c$) or at low values of θ/θ_c, at Reynolds numbers $\mathrm{Re}_* = u_* d/\nu < 22$ to 27, and are said to occur only with fine sediments, $d < 0.7$ to 0.9 mm. The size of the ripples is considered to be independent of flow depth. An empirical formula for average ripple length is

$$\lambda = 1000 d \tag{2.32}$$

and the ripple steepness h/λ as a function of θ/θ_c is shown in Figure 2.8. However, Equation 2.32 is only a rough estimate and the λ-values can differ from this estimate appreciably.

Dunes are generally larger bed features and develop at higher bed shear stresses than ripples. They are the result of the boundary interaction with the flow depth. Their wave

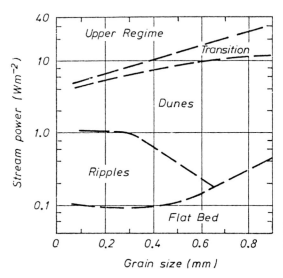

Figure 2.7. Classification of bed features according to Simons et al. (1964).

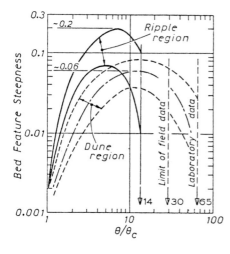

Figure 2.8. Steepness of ripples and dunes according to experimental data. Dune data for $Re_*> 32$ and $y_o/d > 100$, after Yalin (1972).

length is thought to be proportional to water depth. Yalin (1972) argued that

$$\lambda = 2\pi y_o \tag{2.33}$$

The empirical steepness range is shown in Figure 2.8. Führböter (1979) derived for dunes in equilibrium conditions

$$\frac{h}{y_o} = \frac{2}{2n+1} \tag{2.34}$$

where n is the exponent in a relationship for sediment transport q_s as a function of mean velocity, $q_s \approx U^n$, where n appears to be within $3 < n < 6$.

With increasing bed shear stress, the dunes become longer and flatter, leading at times, to what is described as an *undulating bed*, and this gives place to *transition flat bed*. At still higher shear stresses, *antidunes* develop, which are closely linked with surface waves. Antidunes are not common in nature.

The subject of bed features has been the centre of attention for a long time but many questions remain unanswered. For details, reference is made to Yalin (1972), Graf (1971), Raudkivi (1976, 1982, 1990), Kennedy (1969), Engelund (1970), Richards (1980), and Fredsøe (1982).

The different treatments of the problem of bed features are confined to uniform bed material. However, dunes formed from a uniform gravel differ markedly from those on a bed composed of a broad grading of sediment sizes, e.g. for $\sigma_g = 3.5$. Also, where flood flows bring in heavy loads of colloidal suspended sediment, the bed features are affected. Their development is delayed, they are lower and the transition flat bed appears sooner than in a flow without suspended clay.

For a *gravel bed* the variation of the form drag component is not so pronounced and usually the slope of the total shear stress function remains positive with increasing velocity, i.e. no downward slope such as that shown in Figure 2.6. In gravel rivers the bed features are much more rounded in form and their form drag contribution is usually small compared to energy loss due to surface drag and secondary currents. As a consequence roughness of gravel rivers is more readily described by formulae than that of sandy streams. However, special difficulties arise with the description of roughness in very shallow gravel rivers where the roughness height is of the same order of size as the flow depth. The flow in such shallow streams, usually braided, is also twisting and turning, bifurcating and joining together, and all this consumes energy as well.

Hey (1979) presented a detailed discussion of the resistance to flow in gravel rivers, including the effects of cross-sectional shape and different bank roughnesses, as did Limerinos (1970). Hey proposed

$$\frac{1}{\sqrt{f}} = 2.03 \log \frac{11.75R}{3.5d_{84}} \quad \text{or} \quad \left(\frac{8}{f}\right)^{1/2} = 5.62 \log \frac{aR}{3.5d_{84}} \qquad (2.35)$$

where a varies with slope ($11.1 \leq a \leq 13.46$). This differs from the formula by Limerinos only in the factor 3.5 instead of 3.16. The relationship yields reasonable values when y_o/d_{84} is greater than about 6. The data comes from rivers with slopes less than about 1%. Equation 2.35 tends to define the lower limit of roughness for uniform flows; values over 60% greater are possible (Bathurst 1985). Bathurst proposed with data from rivers with slopes steeper than 0.4%

$$\left(\frac{8}{f}\right)^{1/2} = 5.62 \log \left(\frac{y_o}{d_{84}}\right) + 4 \qquad (2.36)$$

which lies more or less in the middle of the range of the data.

For narrow and shallow gravel rivers Bathurst (1978) proposed from field data,

$$\left(\frac{8}{f}\right)^{1/2} = \left(\frac{R}{0.365d_{84}}\right)^{2.34} \left(\frac{B}{y_o}\right)^{7(\lambda - 0.08)} \qquad (2.37)$$

where B is the width of the stream with depth y_o and $\lambda = 0.139 \log (1.91 d_{84}/R)$.

In steep mountain streams the individual roughness elements are large and may even protrude through the water surface. This makes the concept of roughness height at least questionable. The energy losses are mainly governed by the dominant roughness. Thompson and Campbell (1979) suggested

$$\frac{1}{\sqrt{f}} = 2\left(1 - \frac{bk}{R}\right) \log \frac{12R}{k} \qquad (2.38)$$

From their field results $b = 0.1$ and $k = 4.5d$, where d is the diameter of the stones (median d of a counting sample).

The usual approach is to separate the roughness of an alluvial bed into surface or grain roughness and roughness due to bed features, the form drag. Not even for the grain roughness is there unanimity on which size fraction is the representative one. The recommended values range from $k = 1.25 d_{35}$ (Ackers and White 1973) to $k = 5.1 d_{84}$

(Mahmood 1971). Einstein and Barbarossa (1952) used $k = d_{65}$, Engelund and Hansen (1967) $k = 2d_{65}$, Hey (1979) $k = 3.5d_{65}$ etc.

The total resistance is often expressed in terms of the Manning n. Bruschin (1985) expressed the Manning n for the lower regime of sandy rivers as

$$n_L = \frac{d_{50}^{1/6}}{12.38}\left(\frac{RS}{d_{50}}\right)^{1/7.3} \tag{2.39}$$

and the upper regime as

$$n_U = \frac{d_{50}^{1/6}}{20.38}\left(\frac{RS}{d_{50}}\right)^{1/9} \tag{2.40}$$

i.e. the Manning n is

$$n = \frac{k^{1/6}}{a\sqrt{g}} \tag{2.41}$$

where from above $a_L = 3.95$ and $a_U = 6.5$. Strickler's expression for n is equivalent to $a = 6.74$. According to Bruschin, $4.6 < a < 5.3$ for gravel rivers. These formulae provide first order estimates of the value of Manning n.

For flow over bed features, van Rijn (1982) fitted the following to data from flumes and field

$$\frac{k}{h} = 1.1(1 - e^{-25h/\lambda}) \tag{2.42}$$

where h is the height of the bed forms, and λ their length.

The ranges for the data are $0.08 \leq y_o \leq 0.75$ m, $0.25 \leq U \leq 1.1$ m/s and $0.1 \leq d_{50} \leq 2.4$ mm. Equation 2.42 appears to follow the centre of the band of data points, Figure 2.9. He expressed the total hydraulic roughness of an alluvial bed as

$$k = 3d_{90} + 1.1h(1 - e^{-25h/\lambda}) \tag{2.43}$$

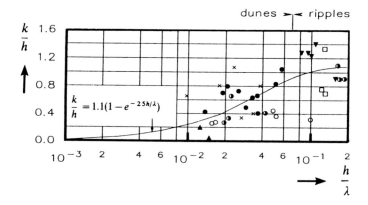

Figure 2.9. Equivalent form drag roughness as a function of height and steepness of bed features, according to van Rijn (1984).

Further complications arise from the time lag in the development of the bed features. Thus, if the flow conditions are such that the bed features increase with increasing flow rate, then with relatively rapidly increasing flow the roughness will be less than that indicated for the instantaneous flow rate. If, however, the flow approaches transition flat bed conditions then the roughness is greater than that for the given flow rate.

The description of the resistance to flow as a function of sediment and flow parameters is one of the unsolved problems of loose boundary hydraulics. The methods of prediction in current use are essentially empirical and have a low level of accuracy. The reasons for this are many. Not even in a steady two-dimensional flow can the roughness height and the dimensions of bed forms be predicted reliably. In nature, the flow rate varies and the bed is covered with residual and newly formed features. The features are usually strongly three-dimensional, the composition of the bed material is not uniform and frequently the material varies over the cross section as well as along the stream.

A very helpful collection of data on river geometry, the measured values of Manning's n and photographs of the rivers was published by the US Geological Survey (Barnes 1967).

The above arguments in general apply to two-dimensional situations. Complications arise even for the side slopes of the flow cross section. For the same bed material the side slope lowers the threshold of movement. Usually the particle size also varies over the periphery. The resistance, particularly over ripples and dunes, is strongly affected by heavy concentrations of colloidal wash load. The development of bed features starts at higher bed shear stresses than in clear water. The features which form are lower and smoother and they are washed out sooner, i.e., the transition flat bed condition occurs at lower shear stresses than in clear water. Significant energy losses arise from momentum transfer in cross sections which have a deep channel flanked with shallow flow areas, for example, over bank flows on flood plains. For these problems reference is made to Knight and Demetriou (1983), Rajaratnam and Ahmadi (1979, 1981).

Natural streams are neither two-dimensional or straight. Frequently sandy streams have cohesive banks and it may be necessary to consider the bank resistance separately. A simple method is to consider the cross section as areas associated with bed and banks $A = A_b + A_w$. Then

$$A_b = \frac{P_b f_b U^2}{8gS} \quad A_w = \frac{P_w f_w U^2}{8gS} \quad A = \frac{PfU^2}{8gS} \tag{2.44}$$

where P is the length of the respective wetted perimeters, and

$$Pf = P_b f_b + P_w f_w \tag{2.45}$$

In meandering streams, bend losses have to be added to friction loss. Rouse (1965) from laboratory tests with 90° bends gave the loss as $0.5U^2/2g$ for $B/y_o = 16$ and $0.2U^2/2g$ for $B/y_o = 8$. However, losses vary with curvature, length of arc and the Froude number. A rough average value for rivers is $0.2U^2/2g$. In laboratory channels an approximate relationship for the gradient, S'', in a bend is

$$\frac{S''}{S} = 300 \left(\frac{y_o}{r_c}\right)^2 \tag{2.46}$$

where r_c is the centre line radius. In sharp bends the lee eddies that form downstream of the bend reduce the effective cross section and consume flow energy. Rozovskii (1957) expressed the energy gradient S'' due to secondary currents in the flow through the bend as

$$S'' = \left(\frac{12\sqrt{g}}{C} + \frac{30g}{C^2}\right)\left(\frac{\bar{y}_o}{r_c}\right) R^2 \mathrm{Fr}^2 \tag{2.47}$$

where C is the Chézy coefficient, $\bar{y}_o$ is mean depth and $\mathrm{Fr}^2 = U^2/g\,\bar{y}_o$

Vanoni (1975) described the proposed methods of estimation of flow depths in alluvial channels, and Raudkivi (1982) presented numerical examples. In principle, the methods are of two kinds; a simultaneous solution of the continuity and Chézy or Manning equations, or separation of surface (grain roughness) drag and form drag due to bed features and use of empirical functions. The first group includes the *regime method*, for example, the Lacey formula

$$U_o = 0.646 \sqrt{fR} = 10.8 R^{2/3} S^{1/3} \text{ m/s} \tag{2.48}$$

where R is the hydraulic mean radius, f is the Lacey silt factor and U_o is the stable channel velocity. Both forms are based on the same data, but for general application the data on the silt factor are meagre. In a wide channel $R \approx y_o$ and $q = U_o y_o$, i.e., Equation 2.48 defines the y_o vs U_o function and its intersection with $y_o = q/U_o$ defines the flow depth for the given flow rate q. The regime method has been refined by inclusion of the side and bed factors (Blench 1957, 1969).

Engelund (1966, 1967) divided the resistance into surface and form drag. His calculation of the depth-discharge relationship proceeds as follows:

(1) Assume y'_o (a depth corresponding to grain resistance) and calculate

$$\theta' = \frac{y'_o S}{\Delta d_{50}} \tag{2.49}$$

(Engelund used the mean fall diameter)

(2) From Figure 2.10 or $\theta' = 0.06 + 0.4\theta^2$ for the ripple and dune region, find the dimensionless shear stress θ. For $0.4 < \theta' < 1.0$ two values of θ can be obtained, one for lower and one for upper regime of flow.

(3) From θ calculate

$$y_o = \frac{\theta \Delta d}{S} \tag{2.50}$$

(4) Calculate mean velocity and discharge from

$$\frac{U}{\sqrt{gy'_o S}} = 5.75 \log \frac{y'_o}{2d_{65}} + 6 \quad \text{and} \quad q = U y_o \tag{2.51}$$

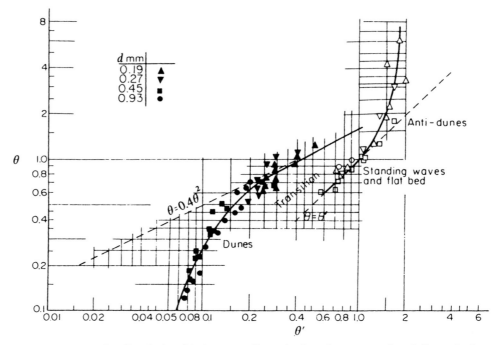

Figure 2.10. Engelund's relationship between dimensionless shear stress θ and dimensionless grain shear stress θ'.

The methods which rely on separation of shear into components due to grain roughness and form drag assume, in principle, that a unique relationship exists between the grain roughness component and total shear. No allowance is made for changes in viscosity or turbulence intensity near the bed. The energy dissipated by form drag increases turbulence which affects sediment suspension, apparent viscosity and size of bed features.

White et al. (1979, 1980) argued that the assumption of $\tau'_o = f(\tau_o)$ is insufficient and that at least the effect of the dimensionless grain size $D_* = [\Delta g d^3/v^2]^{1/3}$ should be accounted for. Data were analysed in terms of the "mobility parameter", F_{gr}, introduced by Ackers and White (1973), and D_*:

$$F_{gr} = \frac{u_*^n}{(\Delta g d)^{1/2}} \left\{ \frac{U}{\sqrt{32} \log(10 y_o/d)} \right\}^{1-n} \tag{2.52}$$

which for fine sediments ($n = 1$) reduces to

$$F_* = \frac{u_*}{(\Delta g d)^{1/2}} = \sqrt{\theta} \tag{2.53}$$

It was found that the data could be fitted by a function of the form

$$\frac{F_{gr}-A}{F_*-A} = f(D_*) = 1 - 0.76\{1 - \exp[-(\log D_*)^{1.7}]\} \qquad (2.54)$$

for Froude numbers less than 0.8, i.e., the lower flow regime, using d_{35} of parent material, or

$$f(D_*) = 1.0 - 0.70\{1.0 - \exp[-1.4(\log D_*)^{2.65}]\} \qquad (2.55)$$

in terms of d_{65} of surface material size distribution. In these equations,

$$\left.\begin{array}{l} n = 0 \\ A = 0.17 \end{array}\right\} \text{ for } D_* \geq 60$$

$$\left.\begin{array}{l} n = 1.0 - 0.56 \log D_* \\ A = (0.23/\sqrt{D_*}) + 0.14 \end{array}\right\} \text{ for } 1 < D_* < 60$$

Thus, if $u_* = (gy_oS)^{1/2}$ and F_* are known, Equation 2.54 or 2.55 yields F_{gr} which in turn, Equation 2.52, yields the mean velocity U and the friction factor

$$f = 8(u_*/U)^2 \qquad (2.56)$$

Van Rijn (1984) proposed a trial and error solution for flow depth, based on Figure 2.9 and the transport parameter T used to describe the height and length of bed features. The procedure is as follows.

(1) Compute dimensionless particle size D_*
(2) Compute u_{*c} from θ_c versus D_*
(3) Compute transport stage $T = [(u'_*)^2 - (u_{*c})^2]/(u_{*c})^2$
where $u'_* = (g^{0.5}/C')U$; $C' = 18\log(12R_b/3d_{90})$
(4) Compute bed form height h from $h/y_o = 0.11(d_{50}/y_o)^{0.3}(1 - e^{-0.5T})(25 - T)$
(5) Compute bed form length λ from $\lambda = 7.3y_o$
(6) Compute equivalent roughness $k = 3d_{90} + 1.1h(1 - e^{-25h/\lambda})$
(7) Compute $C = 18\log(12R_b/k)$

The known values are mean velocity U, flow depth y_o, width, particle size d_{50} and d_{90} as well as density and viscosity. The hydraulic mean radius R_b relates to the bed, i.e., for flume data a side wall correction procedure is applied (e.g. Vanoni and Brooks 1957, Williams 1970).

In practical applications it is prudent not to rely on a particular formula. The selection of formulae or methods should depend on whether or not these were based on data similar to those of the problem at hand. It is also advisable to use more than one formula and compare the results.

2.4. SEDIMENT TRANSPORT

The transport of sediment by water is frequently, for convenience, subdivided into bed *load*, and *suspended load*. A rule of thumb subdivision is

$6 > w/u_* > 2$ bed load
$2 > w/u_* > 0.7$ saltation
$0.7 > w/u_* > 0$ suspension

The saltation phase in water is rather narrow because of the small ratio of solid and water densities. The bed load phase occurs if shear stresses are low. However, once suspension has developed, the subdivision tends to lose physical meaning.

Frequently, another type of transport is present, the *wash load*. The wash load consists of very fine particles, usually clay and fine silt. These particles are brought into the stream by overland flow or bank erosion and are usually not present in quantities in the bed material. An additional source of wash load is the abrasion of gravel in transport. Wash load cannot be computed, because it is not related to local hydraulic parameters.

The various sediment transport relationships could be divided into four groups, according to their derivation. The oldest of these are the relationships in terms of $(\tau_o - \tau_c)$ or $(\theta - \theta_c)$, then come formulae based on probabilistic arguments, formulae based on work done (stream power) and more recently, formulae based on computer optimization of observed data using various dimensionless parameters. Since all the formulae are based on experimental data, the quality of predictions depends on whether the conditions correspond to those on which the formula was based. The assessment of the quality of any formula depends not only on the formula but also on the fact that sediment transport in nature is extremely difficult to measure accurately.

The majority of the formulae can be reduced to the form

$$\phi = f(\theta) \tag{2.57}$$

where

$$\phi = \frac{g_B}{g\rho_s}\left\{\frac{\rho}{\rho_s - \rho}\frac{1}{gd^3}\right\}^{1/2} = \frac{q_B}{(\Delta gd^3)^{1/2}}$$

$$\theta = \frac{\tau_o}{\rho g \Delta d}$$

g_B, g_S and g_T are transport rates in weight per unit width of bed load, suspended load and total transport, respectively and q_B, q_S, q_T the above in material volume rate of transport per unit width. The bulk volumes are obtained by dividing the q-terms by $(1-n)$, where n is the porosity of the deposit.

The formulae are used with a characteristic grain size, or the sediment grain-size distribution is subdivided into a histogram and the results per fraction are added. Since the bed roughness (bed features) varies with shear stress and only part of the latter is used to transport sediment, most formulae attempt to account for this by inclusion of a correction parameter (roughness factor, ripple factor). If one followed the partition

$$\tau_o = \tau_o' + \tau_o'' \tag{2.58}$$

or

$$u_*^2 = u_*'^2 + u_*''^2 \tag{2.59}$$

or

$$y_o S = (y_o S)' + (y_o S)'' \tag{2.60}$$

where τ_o' is shear due to grain roughness of a flat surface (causing grains to move) and τ_o'' shear due to bed features (form drag), then the correction factor could be expressed as

$$r = \frac{(y_o S)'}{y S} = \left(\frac{C}{C'}\right)^2 \tag{2.61}$$

where C is the Chezy coefficient.

Here, only a few of the transport formulae are briefly introduced:

Meyer-Peter and Müller formula (1948)

$$\frac{\gamma R (k/k')^{3/2} S}{d} - 0.047(\gamma_s - \gamma) = 0.25 \rho^{1/3} \frac{(g_B')^{2/3}}{d} \tag{2.62}$$

where $\gamma RS = \rho u_*^2$ and empirically $(k/k')^{3/2} = r$ instead of $S'/S = (k/k')^2$ as by Equation 2.61, g_B' is the immersed weight of sediment and $k' = 26/(d_{90})^{1/6}$. The experiments were carried out with coarse material and relate only to bed load.

The formula can be written as

$$\phi = 8(\theta - 0.047)^{3/2} \tag{2.63}$$

or in terms of bulk volume

$$\phi = \frac{8}{1-n}(\theta - 0.047)^{3/2} \tag{2.64}$$

Engelund-Hansen formula (1967) predicts total load q_T from

$$f \phi = 0.1 \theta^{5/2}; \quad f = \frac{2 g y_o S}{U^2} \quad \text{or} \quad f = \frac{2g}{C^2} \tag{2.65}$$

where C is the Chezy coefficient. The formula is based on experiments with uniform sands $d > 0.19$ mm and $\sigma_g < 1.6$. It has been found to give good results where suspended sediment load is high, particularly if C is based on actual observations.

Ackers and White (1973) proposed a transport function, based on dimensionless terms, for the total load

$$F_{gr} = \frac{u_*^n}{(g \Delta d)^{1/2}} \left\{ \frac{U}{\sqrt{32} \log \alpha y_o/d} \right\}^{1-n} \tag{2.66}$$

$$D_* = d(g \Delta/v^2)^{1/3}; \quad d = d_{35} \tag{2.67}$$

$$G_{gr} = (q_T/Ud)(u_*/U)^n \tag{2.68}$$

which were linked by

$$G_{gr} = C[(F_{gr}/A - 1)]^m \tag{2.69}$$

where the coefficients A, C, m and n were determined from experimental data in terms of D_* using a numerical optimization technique, and $\Delta = (\rho_s - \rho)/\rho$. The exponent $n \to 0$ for coarse sediments ($D_* > 60$ or $d > 2.5$ mm) and $n \to 1$ for fine sediments ($D_* < 1$ or $d < 0.04$ mm). The coefficient α in turbulent rough boundary flow was approximated by $\alpha = 10$ and the fitted functions for n, m, A and C are as follows:

(a) Coarse sediments $D_* > 60$: $n = 0$; $A = 0.17$; $m = 1.5$ and $C = 0.025$.
(b) Transition range $60 > D_* > 1$:

$$n = 1.00 - 0.56 \log D_*$$
$$m = (9.66/D_*) + 1.34$$
$$A = (0.23/\sqrt{D_*}) + 0.14$$
$$C = 10^{[2.86 \log D_* - (\log D_*)^2 - 3.53]}$$

or

$$\log C = 2.86 \log D_* - (\log D_*)^2 - 3.53.$$

The value of A represents a critical value of the mobility number F_{gr}, the threshold condition $(\theta_c)^{1/2}$. For $1 < D_* < 60$, the Shields curve is approximated by the transition function. For coarse grains θ_c is set at $(0.17)^2 = 0.029$ which is appreciably lower than the usual values.

With the aid of data from over 1000 flow experiments and 260 field measurements, White et al. (1975) tested the method against seven other formulae. They showed that 68% of data points fell within the range $0.5 < q_{cals}/q_{obs} < 2$, compared to 63% for Equation 2.65, 46% for the Einstein method, etc.

Suspension is a very important mode of transport of sediment since the quantities transported in suspension are usually very much larger than those in bed load. Although the methods of description of suspensions are much more advanced than those of bed load, these still yield only the distribution of sediment concentration with depth, given a reference concentration. It is not yet possible to compute the suspended sediment transport capacity of a given flow.

The analytical models of suspension may be subdivided into diffusion, energy and stochastic models of which the diffusion model is most widely used.

The diffusion model of suspension is based on an analogy between turbulent diffusion and diffusion in laminar flow described by

$$\frac{\partial c}{\partial t} + u_i \frac{\partial c}{\partial x_i} = D \frac{\partial^2 c}{\partial x_i \partial x_i} \tag{2.70}$$

where D is the diffusion coefficient in laminar flow.

The analogy for turbulent flow leads to

$$\frac{\partial \bar{c}}{\partial t} + \bar{u}_i \frac{\partial \bar{c}}{\partial x_i} = -\frac{\partial}{\partial x_i}\left(D_{ij}\frac{\partial \bar{c}}{\partial x_j} + D_{ji}\frac{\partial \bar{c}}{\partial x_i}\right) \tag{2.71}$$

The coefficient of turbulent diffusion D_{ij} is a second order tensor and its values are a function of space, whereas the laminar diffusion coefficient is a scalar quantity. Generally D_{ij} is much larger than D so that D can be neglected. If the turbulence is *homogeneous*, D_{ij} reduces to D_{ii} (D_{xx}, D_{yy}, D_{zz}) and if the turbulence is *isotropic*, D_{ij} reduces to a scalar $D_T = \varepsilon_s$. Then Equation 2.71 becomes for steady flow

$$w\frac{dc}{dy} + \frac{d}{dy}\left(\varepsilon_s \frac{dc}{dy}\right) = 0 \tag{2.72}$$

where ε_s is the sediment diffusion coefficient. Integrating once yields

$$wc + \varepsilon_s \frac{dc}{dy} + A = 0 \tag{2.73}$$

where the constant of integration A is zero if the concentration at the water surface is zero. Rouse (1937) assumed a logarithmic velocity distribution, a linear shear stress variation and $\varepsilon_s = \varepsilon$, the momentum exchange coefficient (eddy viscosity). With these he obtained a solution of Equation 2.73 for $A = 0$ of the form

$$\frac{c}{c_a} = \left[\left(\frac{y_0 - y}{y_0 - a}\right)\frac{a}{y}\right]^z \tag{2.74}$$

where

$$z = \frac{w}{\kappa u_*}$$

κ is the Karman constant and c_a is a reference concentration at depth $y = a$, a small distance above the bed. Notwithstanding all the assumptions involved, the equation predicts the distribution of sediment, particularly fine sediments, quite well. Many refinements have been proposed but these yield little compared with the difficulties faced with definition of ε_s.

The ε_s distribution, modelled after ε, is parabolic with zero values at the bed and the surface. Measurements by Coleman (1970) show that ε_s remains essentially constant from about 0.2 y_0 up. These results also show that coarser sediment has larger ε_s values than fine sediment. A likely reason for this is that the larger grains acquire enough momentum to fly out of the eddy system of water turbulence and sample the larger eddies of turbulence of neighbouring systems whereas the fine particles are more trapped within the eddy system of the water movements.

Generally Equation 2.74 can be fitted to data quite well by using an appropriate z-value. This led van Rijn (1984) to express z as $z_1 = w/(\beta \kappa u_*)$ where empirically

$$\beta = 1 + 2(w/u_*)^2; \quad 0.1 < w/u_* < 1 \tag{2.75}$$

His empirical expression for c_a is

$$c_a = 0.015(d_{50}/a)T^{1.5}/D_*^{0.3}; \quad T = (u_*'^2 - u_{*c}^2)/u_{*c}^2$$

where $a \approx k$, the roughness height; $D_* = (\Delta g d^3/v^2)^{1/3}$; $u_*' = (U\sqrt{g})/C'$ and C' is Chezy coefficient based on grain roughness.

Engineering calculations for particular rivers are aided if data on c_a are accumulated and c_a plotted as a function of flow rate. However, for rivers with measurements of suspended sediment and flow rate one can prepare an empirical function of suspended sediment load versus flow rate and thus avoid the calculation.

Only nearly straight river reaches with well shaped channels produce flows with approximately homogeneous turbulence and the textbook type distribution of sediment. The usual river contains significant secondary (spiral) currents and a multitude of flow features, which are frequently referred to as macro-turbulence. Of the latter, continuous rotating movements, such as eddies with vertical axes, and concentrated vortices have a strong influence on suspended load distribution. Vortices in the surfaces of discontinuities, in the lee of bed features etc. can entrain significant amounts of sediment. Lee vortices which lift off from the bed, and form into vortex rings, traverse the depth and appear on the surface as "boils". Measured distributions of suspended sediment concentrations in these regions tend to show not only higher concentrations but also different values of the exponent z. Special problems occur when the flow carries very high concentrations of sediment (e.g. the Yellow River and mud flows or lahars).

A feature, which has caused a great deal of confusion is the ε-value in a turbulent suspension. Vanoni and Nomicos (1959) determined the ε-value from logarithmic velocity distributions. This could be misleading since this distribution is valid only in the lower 15–20% of depth, and data are usually obtained from the upper 70%. Coleman (1981, 1985) and Raudkivi (1982) have discussed this point.

The *suspended sediment transport* is given by

$$q_s = \int_{y=a}^{y=y_o} cu \, dy \tag{2.76}$$

Einstein (1950) combined this with the logarithmic velocity distribution

$$q_s = \int_{y=a}^{y=y_o} c_a \left\{ \frac{y_o - y}{y} \frac{a}{y_o - a} \right\}^z 5.75 u_* \log(30.2 y/k) dy$$

$$= 11.6 u_* c_a a [2.303 \log(30.2 y_o/k) I_1 + I_2] \tag{2.77}$$

where the integrals I_1 and I_2 are presented in graphical and tabular form and k is the apparent roughness. With a further assumption that $a = 2d$ and that $c_a = i_B q_B / 11.6 u_* a$, where $i_B q_B$ is the fraction of bed load in the grain size class i

$$i_T q_T = i_B q_B [2.303 \log(30.2 y_o/k) I_1 + I_2 + 1] \tag{2.78}$$

All the transport formulae have been developed for straight channels, and little is known about how these should be applied to a meandering river. Since the transport relationships are non-linear, the use of average parameters is likely to be unsatisfactory. If it is feasible to delineate streamlines adequately, the width of the river can be divided into constant depth bands and the transport summed.

The formulae also imply steady state conditions. Fortunately, the variation of stream flows is usually gradual enough to allow computation of transport rates as steady state increments.

A major problem is the calculation of transport rate in gravel rivers. Only when the shear velocity exceeds that of the limiting armour layer, u_{*a}, does a well defined steady sediment transport take place. At $u_* < u_{*a}$ the bed slowly armours and sediment transport rate goes to zero. When the bed has armoured itself for conditions near u_{*a}, then almost no transport of bed material occurs for all flows with $u_* \leq u_{*a}$, except for that brought in by tributaries which moves over the armoured bed.

The most difficult problems arise when the sediment supply changes, for example, when the sediment is intercepted in a reservoir. Then, if the river flow continues downstream as for hydro-power reservoirs, a gradual degradation of the river-bed below the reservoir takes place. The earlier stages of this process can be modelled numerically as discussed by Jansen et al. (1979). The latter stages of the process involve also the effects of armouring; that is, profile adjustments occur only when bed shear stresses exceed those of the largest grain fraction in the bed sediment.

REFERENCES

Ackers, P. & W.R. White 1973. Sediment transport: new approach and analysis. *Proc. ASCE* 99(HY11); 2041–2061.
Allen, J.R.L. 1965. A review of the origin and characteristics of recent alluvial sediments. *Sedimentology*, Vol. 5; 89–191.
Barnes, H.H. Jr. 1967. *Roughness characteristics of natural channels.* US Geological Survey, Water-Supply Paper 1849.
Bathurst, J.C. 1978. Flow resistance of large-scale roughness. *Proc. ASCE* 104(HY12); 1587–1603.
Bathurst, J.C. 1985. Flow resistance estimation in mountain rivers, *J. Hydr. Eng., ASCE*, 111(4); 625–643.
Blench, T. 1957. *Regime behaviour of canals and rivers.* Butterworth, London.
Blench, T. 1969. *Mobile-bed fluviology.* University of Alberta Press, Canada (first edition 1966).
Bogardi, J.L. 1974. *Sediment transport in alluvial streams.* Budapest, Akadémiai Kiado.
Bruschin, J. 1985. Disc. Flow depth in sand-bed channels. *J. Hydr. Eng., ASCE* 111; 736–739.
Chin, C.O. 1985. *Stream bed armouring.* Univ. of Auckland, N.Z., Dept. of Civil Engng. Rep. No. 403.
Christensen, B.A. 1969. Effective grain size in sediment transport. *Proc. 13th Congress IAHR, Kyoto,* 3; 223–231.
Coleman, N.L. 1970. Flume studies of the sediment transfer coefficient. *Water Resources Res.* 6; 801–809.
Coleman, N.L. 1981. Velocity profiles with suspended sediment. *J. Hydr. Res.* 19; 211–230.
Coleman, N.L. 1985. Effects of suspended sediment on open-channel velocity distributions.

Euromech 192, Munich/Neubiberg, F.R. Germany.
Einstein, H.A. 1950. *The bedload function for sediment transport in open channel flows.* U.S. Dept. Agricult., Soil Conservation Service. Techn. Bull. No. 1026.
Einstein, H.A. & N.L. Barbarossa 1952. River channel roughness. *Proc. ASCE* 77. Separate No. 78, *Trans. ASCE* 117; 1121–1146.
Engelund, F. 1966. Hydraulic resistance of alluvial streams. *Proc. ASCE* 92(HY2; 315–326.
Engelund, F. 1967. Closure of discussion. *Proc. ASCE* 93(HY4); 287–296.
Engelund, F. 1970. Instability of erodible beds. *J. Fluid Mech.* 42; 225–244.
Engelund, F. & J. Fredsøe 1982. Sediment ripples and dunes. *Ann. Rev. Fluid Mech.* 14; 13–37.
Engelund, F. & E. Hansen 1967. *A monograph on sediment transport in alluvial streams.* Teknisk Vorlag, Copenhagen.
Fenton, J.D. & J.E. Abbott 1977. Initial movement of grains on a stream bed: the effect of relative protrusion. *Proc. Roy. Soc. London*, A352; 523–537.
Fredsøe, J. 1982. Shape and dimensions of stationary dunes in rivers. *Proc. ASCE* 108(HY8); 932–947.
Führböter, A. 1979. Strombänke (Grossriffel) und Dünen als Stabilisierungsformen. *Mitt. Leichtweiss-Institut. TU Braunschweig* 67; 155–192.
Garde, R.J. & K.G. Ranga Raju 1977. *Mechanics of sediment transport and alluvial stream problems.* Wiley Eastern, Delhi. Second Edition 1985.
Graf, W.H. 1971. *Hydraulics of sediment transport.* McGraw-Hill, New York.
Hey, R.D. 1979. Flow resistance in gravel-bed rivers. *Proc. ASCE* 105(HY4); 365–379.
Hjülström, F. 1935. The morphological activity of rivers as illustrated by River Fyris. *Bull. Geol. Inst., Uppsala* 25, Chap. III.
Irvine, M.H. & A.J. Sutherland 1973. A probabilistic approach to the initiation of movement of non-cohesive sediments. *Proc. IAHR Int. Symposium on River Mechanics, Bangkok*, 1; 383–394.
Jansen, P.Ph. (ed.) 1979. *Principles of River Engineering.* The non-tidal alluvial river. Pitman, London.
Kennedy, J.F. 1969. The formation of sediment ripples, dunes and anti-dunes. *Ann. Rev. Fluid Mech.* 1; 147–168.
Knight, D.W. & J.D. Demetriou 1983. Flood plain and main channel flow interaction. *J. Hydr. Eng., ASCE* 109; 1073–1092.
Lacey, G. 1929. Stable channels in alluvium. *Proc. Inst. Civil Engrs.* 229; 259–384.
Leopold, L.B., G.M. Wolman & J.P. Miller 1964. *Fluvial process in geomorphology.* W.H. Freeman and Co., San Francisco.
Limerinos, J.T. 1970. *Determination of the Manning coefficient for measured bed roughness in natural channels.* U.S. Geological Survey, Water Supply Paper 1898-B.
Mahmood, K. 1971. *Flow in sandbed channels.* Colorado State Univ., Fort Collins, Water Management Tech. Rep. No. 11.
Meyer-Peter, E. & R. Müller 1948. Formulas for bed-load transport. *Proc. 3rd Congress IAHR, Stockholm*; 39–64.
Neill, C.R. 1967. Mean velocity criterion for scour of coarse uniform bed material. *Proc. 12th Congress IAHR, Fort Collins*, 3; 46–54.
Neill, C.R. 1968. *A re-examination of the beginning of movement for coarse granular bed materials.* Rep. INT 68, Hydraulic Research Station, Wallingford.
Nordin, C.F. Jr. 1964. *Aspects of flow resistance and sediment transport Rio Grande near Bernalillo, New Mexico*, U.S. Geological Survey Water Supply Paper 1498-H; also in *Proc. ASCE* 97(1971); 101–141.
Rajaratnam, N. & R.M. Ahmadi 1979. Interaction between main channel and flood plain flows, *Proc. ASCE* 105(HY5); 573–588.

Rajaratnam, N. & R.M. Ahmadi 1981. Hydraulics of channels with flood plains, *J. Hydr. Res.* 19; 43–59.
Raudkivi, A.J. 1967, 1976, 1990. *Loose boundary hydraulics*. 1st, 2nd and 3rd Edition, Pergamon Press, Oxford.
Raudkivi, A.J. 1982. *Grundlagen des Sedimenttransports*. Springer-Verlag, Berlin.
Raudkivi, A.J. & S.K. Tan 1984. Erosion of cohesive soils. *J. Hydr. Res.* 22; 217–233.
Reineck, H.E. & I.B. Singh 1973. *Depositional sedimentary environments*. Springer-Verlag, Berlin. 2nd Edition, 1980.
Richards, K.J. 1980. The formation of ripples and dunes on an erodible bed. *J. Fluid Mech.* 99; 597–618.
Richards, K.J. 1982. *Rivers*, Methuen, London.
Rouse, H. 1937. Modern conceptions of mechanics of turbulence. *Trans. ASCE* 102; 463–543.
Rouse, H. 1965. Critical analysis of open-channel resistance. *Proc. ASCE* 91(HY4); 1–25.
Rozovskii, I.L. 1961. *Flow of water in bends of open channels*. Academy of Science of Ukrainian SSR, Kiev 1957. Translated U.S. Dept. of Commerce No. OTS 60-51133.
Rijn, L.C. van 1982. Equivalent roughness of alluvial bed. *Proc. ASCE* 108(HY10); 1215–1218. Also as: The prediction of bed forms, alluvial roughness and sediment transport. *Delft Hydraulics Laboratory*, Report S 487-III and van Rijn (1984).
Rijn, L.C. van 1984. Sediment Transport Part I: Bed load transport, *J. Hydr. Eng., ASCE* 110; 1431–1456. Part II: Suspended load transport, *J. Hydr. Eng., ASCE* 110; 1613–1641. Part III: Bed forms and alluvial roughness, *J. Hydr. Eng., ASCE*, 110; 1733–1754.
Simons, D.B., E.V. Richardson & W.L. Haushild 1962. Depth-discharge relations in alluvial channels. *Proc. ASCE* 88(HY5); 57–72.
Simons, D.B., E.V. Richardson & C.F. Nordin Jr. 1964. *Sedimentary structures generated by flow in alluvial channels*. Colorado State University, Fort Collins, Rep. CER 64, Am. Assoc. of Petroleum Geologists Special Publication No. 12.
Simons, D.B. & F. Sentürk 1976. *Sediment transport technology*. Water Resources Publications, Fort Collins.
Strickler, A. 1923. Beiträge zur Frage der Geschwindigkeitsformel und der Rauhigkeitszahlen für Ströme, Kanäle und geschlossene Letungen. *Mitteilungen des Eidgenössischen Amtes für Wasserwirtschaft*, Bern, Schwitzerland, No. 16.
Thompson, S.M. & P.L. Campbell 1979. Hydraulics of a large channel paved with boulders. *J. Hydr. Res.* 17; 341–354.
US Interagency Committee 1957. *Some Fundamentals of Particle Size Analysis, A study of methods used in measurement and analysis of sediment loads in streams*. Subcommittee of Sedimentation, Interagency Committee on Water Resources, Report No. 12, St. Anthony Falls Hydraulic Laboratory, Minneapolis, Minnesota.
Vanoni, V.A. & N.H. Brooks 1957. *Laboratory studies of the roughness and suspended load of alluvial streams*. Sedimentation Laboratory, California Institute of Technology, Rep. E-68.
Vanoni, V.A. & G.N. Nomicos 1959. Resistance properties of sediment-laden streams. *Proc. ASCE* 85(HY5); 1170–1173.
Vanoni, V.A. (ed.) 1975. *Sedimentation Engineering*. ASCE-Manuals and Reports on Engineering Practice No. 54. ASCE New York.
White, W.R., H. Milli & A.D. Crabbe 1975. Sediment transport theories: a review. *Proc. Inst. Civil Engrs (London)*, Part 2; 59; 265–292.
White, W.R., E. Paris & R. Bettess 1979. *A new general method for predicting the frictional characteristics of alluvial streams*. Hydraulics Research Station, Wallingford. Rep. No. IT187.
White, W.R. E. Paris & R. Bettess, 1980. The frictional characteristics of alluvial streams: a new approach. *Proc. Inst. Civil Engrs (London)*, Part 2; 69; 735–750.

White, W.R. & T.J. Day 1982. Transport of graded gravel bed material. In: *Gravel-Bed Rivers*, R.D. Hey et al. (eds). Wiley, New York.
Williams, G.P. 1970. *Flume width and water depth effects in sediment transport experiments.* U.S. Geological Survey, Professional Paper 562-H.
Yalin, M.S. 1972. *Mechanics of sediment transport.* Pergamon Press, New York.

CHAPTER 3

Scour in rivers and river constrictions

H.N.C. BREUSERS

3.1. INTRODUCTION

An understanding of the behaviour of river beds and banks is important to the designer of any river structure. Difficult questions need consideration: will upstream or downstream works affect the local river; is there a significant lowering of the bed during a flood; does a localised channel develop; is the bed fluidised so that its bearing capacity is greatly reduced; is the river likely to move laterally during the life time of any proposed structure? Because no general or conclusive answers to these questions are at present available, the designer must depend on experiments with only limited applicability and gain guidance from a review of local observations and the recent evolution of the given river.

The process of scouring in rivers can result from natural phenomena or from man-made alterations; either of these can produce effects over long reaches of the river or only locally. In addition to the extended effects of the natural river regime, local scouring can occur at bends and confluences.

Some rivers scour very little during a flood, whereas others lower their beds substantially. If the flood flows are confined laterally, large changes can occur in the elevation of both the water surface and the bed. If, in contrast, the flood can readily spill out into flat land adjacent to the river the elevations of both water surface and river bed may change comparatively little. In similar ways, man-made changes can alter the supply of sediment or the carrying capacity of a river so as to produce extensive changes in a river bed.

3.2. GENERAL SCOUR

General scour occurs because of the increased capacity of a river to carry sediment during flood flows or as the consequence of various man-made alterations. Some examples of the latter are given by Simons and Sentürk (1977) and Jansen (1979):

— erosion downstream of reservoirs due to the interception of sediment,

- transfer of water from one basin to another which alters the sediment carrying capacities of both rivers,
- mining of sand and gravel in rivers and on flood plains,
- cut-off of meanders,
- decrease in effective boundary roughness and cross-section through channel regulation.

This type of scour is a response to the flow regime, and it would occur whether a structure like a bridge is built or not.

In the process of general scour during floods, the bed of the river is lowered, and the extent to which this occurs is important to the designer. If the elevation and slope of the water surface are known, one can estimate the mean velocity and cross-sectional area for a given discharge. These in turn enable an estimate of the average elevation of the bottom and the amount of scour that this implies. The results of such a simplified approach are only estimates.

Many rivers, especially wide rivers flowing over gravel, have a tendency to develop localised channels as shown in Figure 3.1. The lowest part of the bed in such a channel can be well below the mean bed level.

If this part of the channel should coincide with the location of a bridge pier, the two effects would combine to produce even greater depths of scour. Gravel rivers also contain large moving banks as features of their beds which can produce at least partial bifurcation of the flow and the development of preferred channels. Such networks of

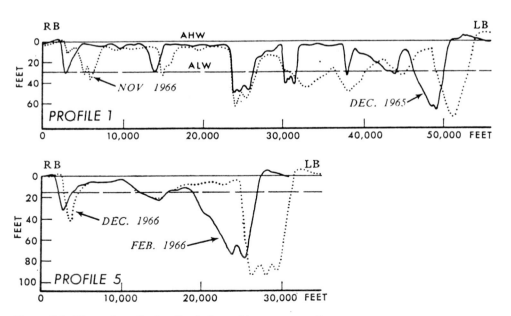

Figure 3.1. Illustration of a localized channel in a cross-section.

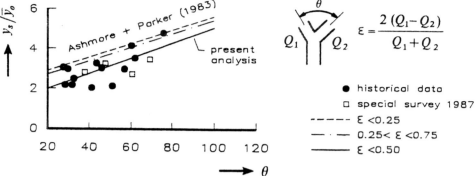

Figure 3.2. Confluence scour Jamuna River.

channels, usually present at low discharges, can thus contribute to marked irregulaties of the bed for high flows as well.

If two streams come together or bifurcated channels reunite, they often do so at an angle and at different levels. The resulting spiral motion can be all too effective in producing a region of significant scouring.

Field data were collected by Ashmore and Parker (1983) for gravel rivers and by Klaassen and Vermeer (1988) for the Jamuna River, Bangladesh (see Figure 3.2).

The values for the Jamuna were slightly lower on the average. The best-fit relation was given by Klaassen and Vermeer as:

$$r\bar{y}_o = 1.29 + 0.037\theta \tag{3.1}$$

where θ is the angle between the two channels, $\bar{y}_o$ the average depth of the two channels and $r\bar{y}_o$ is the maximum depth at the confluence.

Information on possible river bed changes can be obtained for certain conditions from one-dimensional mathematical river models such as HEC-6 (Thomas 1977), RIVMOR (Klaassen et al. 1982) and the Sogreah model (Cunge et al. 1980). For example, Figure 3.3 illustrates predictions made for the consequences of a meander cut-off. These models are limited however because they are one-dimensional and provide for only highly schematized representations of bank erosion and bed-material gradation.

Field studies on the Rio Apure showed dramatic changes in the maximum depth during the passage of a flood. The cross sections of the Rio Apure indicate the water surface and bed profiles during and after the annual flood (Figure 3.4). Near the end of the high flow in 1969, the surface level was 45 m and the maximum elevation of the bottom had degraded to an elevation of 22 m, with a maximum water depth of 23 m. During the subsequent period of low flow, the surface dropped about 6 m and the bottom rose to levels between 32 and 37 m. Thus a tremendous amount of material was scoured from the bottom during the high flow and presumably dropped at locations downstream.

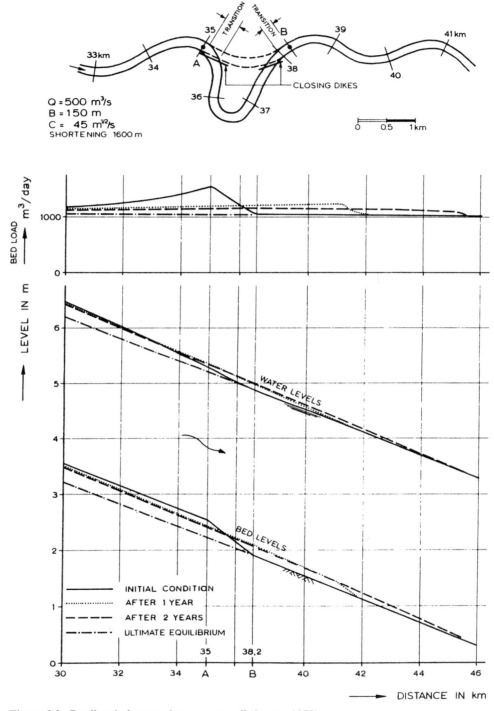

Figure 3.3. Predicted changes due to a cut-off (Jansen 1979).

Scour in rivers and river constrictions 41

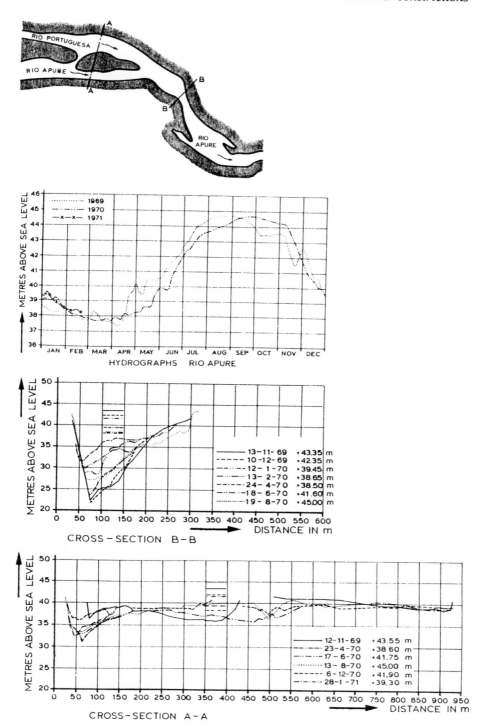

Figure 3.4. Bed level changes Rio Apure (Jansen 1979).

3.3. SCOUR AT BENDS

Lateral movements of unregulated rivers can be as large as 1000 m/year (Kosi River, India, Joglekar (1971)), and even less active rivers move on the order of 10–100 m/year (Schumm, 1977, Shen, 1979). Braided streams are probably the most unstable. Nevertheless even comparatively stable meandering streams can move significant lateral distances over the number of years assigned as the life of a structure. These tendencies can be noted from field observations, maps and aerial photographs.

Both the pattern of flow at a bend and the scour it produces are complicated by secondary flows and by gradation of the sediment. Neither the water surface nor the bed elevation are constant across a bend, and both are difficult to predict. They depend not only upon the nature of the bend (plan shape, bank erodibility, segregation of bed material), but also on lateral water surface slope and a spiral (secondary) flow that develops there. No theory can describe completely such a complex occurrence, but various approximations have been presented. The water surface rises towards the outside of the bends in accordance with the equations of motions, and the bed tends to scour more deeply there due to the spiral flows. Odgaard (1981, 1982, 1984) summarized the available information on scouring at bends and expressed the cross slope of the bed, β, by means of the equation

$$\sin \beta = K(Fr)^\alpha (y_o/r) \qquad (3.2)$$

in which K is a constant, Fr is $U/\sqrt{gy_o}$ and y_o/r is the ratio of the mean depth to the radius of the bend. Although other values for the exponent α have been proposed (Engelund 1974 and Van Bendegom 1947), he concluded that a value of unity represents most effectively the available information from laboratory and field studies. Thus the slope is affected by the velocity of the flow, the relative depth and the sediment size.

Recent developments in numerical modelling are promising. Struiksma et al. (1985) have developed a morphological model of a river bend which predicts the bed levels for a river bend with fixed boundaries. The longitudinal profiles near the inside of the bend (right bank) and the outside of a bend are shown in Figure 3.5 as predicted and as observed. The comparison shows good quantitative agreement, and the model predicts the irregular pattern of scour and fill.

3.4. CONSTRICTION SCOUR

In order to reduce the construction cost of bridges, the cross section at a crossing is often reduced by encroachment of the embankments on the flood plain or even on the river channel. If the bed of the river is mobile, the concentration of flow at the constriction will result in an increase in flow depth through scouring.

The increase in depth at a long constriction can be computed from the equations of motion and continuity for sediment and water. For the water motion the Chézy

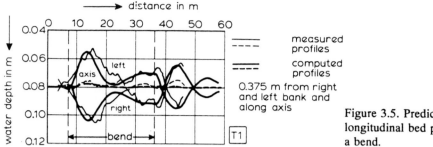

Figure 3.5. Prediction of longitudinal bed profiles in a bend.

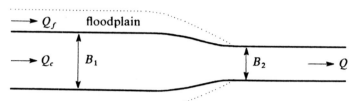

Figure 3.6. Definition sketch for constriction scour.

equation is used and for the sediment transport a simple relationship to the average velocity (for definitions, see Figure 3.6):

water: $\quad Q = \text{constant} = Q_c + Q_f$ (3.3)

sediment: $\quad Q_s = \text{constant} = Bq_t = B(aU^m)$ (3.4)

Neglecting floodplain flow, one finds

$$\frac{y_{o2}}{y_{o1}} = \left(\frac{B_1}{B_2}\right)^\alpha$$ (3.5)

where $\alpha = (m-1)/m$ or with $m = 3$ to 5, $\alpha = 0.67$ to 0.8. The value of $\alpha = 2/3$ corresponds to the value predicted by the regime method. Straub, see Vanoni (1975) and Laursen (1960) give values close to $\alpha = 0.64$. In case of overbank flow with discharge Q_f, Equation 3.5 becomes

$$\frac{y_{o2}}{y_{o1}} = \left(\frac{B_1}{B_2}\right)^\alpha \frac{Q}{(Q-Q_f)}$$ (3.6)

Experimental data to confirm the foregoing analysis are scarce. Sharma (1972) concluded from field data that the scour depth between bridge piers (excluding local scour effects) is reasonably predicted by the regime method. Norman (1975) concluded from field data that either the Straub, or the Laursen approach gives good results.

Laboratory data, Gill (1981) and Webby (1984), indicated that the Laursen approach underestimates the depth y_{o2} in the long contraction by about 20%. If the contraction is rather abrupt, *local* scour can occur like that at the head of a groyne or an abutment. The depth of the local scour depends on the shape of the entrance. Webby quotes $y'_{o2} = 1.5 y_{o2}$ for a streamlined entrance to a contraction and $y'_{o2} = 2 \cdot 6 y_{o2}$ for a blunt entrance, where the local scour depth $y_s = y'_{o2} - y_{o2}$.

In rivers the increase in depth will not be uniform over the width. Neill (1975) suggested a graphical method for estimation of the cross-sectional shape (see Figure 3.7) but this yields little more than a warning to allow for extra scour depth as a safety margin. Actual design should be based on experience with the river under study or from similar situations.

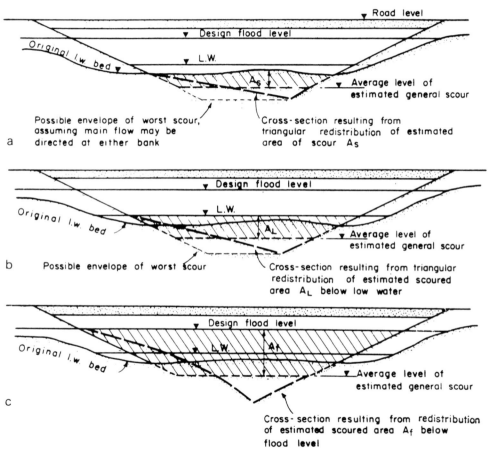

Figure 3.7. Various alternatives in graphically redistributing the cross-sectional area of a scoured waterway opening: (a) redistribution of a net area of scour only (non-alluvial streams with erodible beds); (b) redistribution of scoured area below the low water level (semi-alluvial streams with limited bed movement); (c) redistribution of scoured area below the design flood level (alluvial streams with highly mobile channels).

3.5. LARGE SCALE PLAN FEATURES OF RIVERS

Where rivers are not confined by man-made bank protections, they tend to develop a number of characteristic shapes in planform. On emerging from the mountains, where topography controls the river geometry, a river develops into a pattern of multiple channels, called *braided*. In the lower reaches of the river, a single winding main channel pattern forms, called *meandering*. Leopold et al. (1964) related the slope, S, to the bank full discharge Q_b, which separates the braiding and meandering regimes as:

$$S = 0.0125 Q_b^{-0.44} \qquad (3.7)$$

Rivers with S, Q_b combinations above the line will be in the domain of braiding.

The methods of analysis of meandering are empirical or analytical models based on stability analysis. These are reviewed by Callander (1978) and discussed in numerous texts, Richards (1982). In general, meander patterns show an unmistakable regularity, yet their wave length and amplitude can be described only in statistical sense. The random element arises mainly from the inhomogenity of the terrain through which the river flows. Many relationships have been proposed for meander dimensions. For a definition of meander dimensions, see Figure 3.8.

As an average for all empirical relations the following values hold:

$$L = 10W \quad R \approx 2.5W \qquad (3.8)$$

where W is the bank-full width. This can be estimated with the relation given by Lacey for the wetted perimeter P, which is approximately equal to the width.

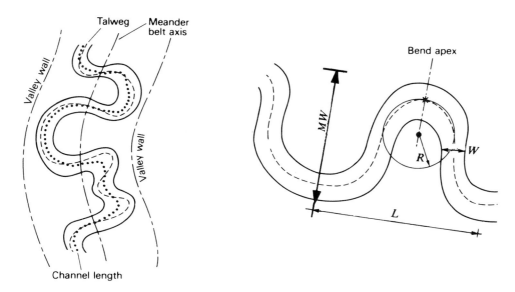

Figure 3.8. Meander dimensions.

$$W \approx P = 4.8 Q_b^{1/2} \tag{3.9}$$

Garde and Ranga Raju (1977) related meander length L and width MW also to Q_b:

$$L \approx (10-12)P \quad MW \approx (18-30)P \tag{3.10}$$

A river can change in character due to regulation, for example by an upstream dam. Schumm (1977) gives as an example the South Platte River, which changed from braided to meandering due to a decrease in flood discharges. River dimensions are also a function of sediment load. If a tributary brings in large quantities of suspended sediment, the width of the river downstream of the confluence may decrease and depth increase. Large bed load input leads to an increased width and decreased depth downstream.

3.6. APPROACH TO DESIGN

The importance of meandering is that the meander patterns propagate slowly downstream. Bends may be cut off by the natural tendency of erosion at the outer banks of bends. A cut-off could for a time destabilize the river in the vicinity of the cut-off because of the locally increased slope. Translation of meander patterns or cut-offs could leave a bridge without a river or lead to a bridge crossing an oxbow lake. Therefore river behaviour and stability should be studied in conjunction with any engineering design.

It cannot be overemphasized that in order to comprehend the likely tendencies of the river to change its shape and location, the designer must collect and study a variety of relevant information including particularly the experience with former constructions in comparable conditions. Otherwise his designs run the risk of encountering a variety of unsatisfactory situations: the bypassing of bridges, the cutting off of abutments, variation in the angle of the approach flow to such structures and serious degradation of the river channel. Such occurrences cannot be eliminated simply by following a set of prescribed rules, but their likelihood of causing serious consequences can be reduced by careful assessment along the lines proposed by Neill (1975). His excellent review outlines important factors to be considered in the selection of a site for a bridge or other structure encroaching the river.

Neill gives the following checklist for basic data of importance in bridge design:

Table 3.1. Checklist for basic data (Neill 1975).

Nature of information	Office data to be assembled or reviewed	Field investigations to be conducted
Maps, charts, airphotos	Topography maps Surficial geology or soils maps Depth charts of navigable waters Small-scale stereo photos for location Large-scale photos for working plans	Look for channel changes since latest maps or photos

Table 3.1. Checklist for basic data (Neill 1975) (continued).

Nature of information	Office data to be assembled or reviewed	Field investigations to be conducted
Existing bridges and other structures (for evaluation of their adequacy and performance)	Dimensions of waterway opening Details and dates of construction, alteration, damage, repairs, failure, etc. Road profile across flood-plain Foundation levels Recorded flood and ice levels at site Subsoil borings Past scour surveys	Check on site Check local evidence and investigate reasons for repairs, etc. Check if road or bridge superstructure has been raised Look for evidence of scour in area of structure, and check adequacy of scour protection, etc. Check local evidence, especially of overflooding or breaching of approaches, or other bypassing of structure by flood waters
Water-level and discharge data	Recorded discharges and stages for nearest hydrometric stations, especially annual maxima Flood-frequency curves for nearest stations or for region Stage discharge curves for nearest stations Flow-duration curves and annual hydrographs Unpublished or unofficial data from various agencies, newspaper files, etc. Records and forecasts for tides, waves, storm surges, etc. in lakes and coastal waters Project floods and related criteria used for upstream dams or related structures	Make thorough search for site evidence on high flood levels: debris, ice scars, stains on structures, witnesses, local photographs, etc. Check if marks were affected by ice-jam or backwater conditions Check credibility of data from office sources Search for local evidence on waves, tides, wind set up, etc. Investigate velocities and directions of maximum currents
Hydraulic geometry and channel capacity	Use airphotos to assist field survey Use airphotos to assist field investigation Check slope from topography maps Check property affected on airphotos	Measure cross-sectional dimensions of channel and flood-plain Photograph channel and adjacent areas Seek evidence of main overflow routes and flood relief channels Measure channel slope Assess property liable to be affected by backwater Search for hydraulic control points: rapids, falls, etc. Assess roughness or flow capacity of flood-plain areas

Table 3.1. Checklist for basic data (Neill 1975) (continued).

Nature of information	Office data to be assembled or reviewed	Field investigations to be conducted
Ice conditions and debris	Recorded ice thickness Recorded dates of break-up and freeze-up	Make thorough search for local evidence on high ice conditions and reasons for their occurrence; on damage to structures; and on thicknesses and dimensions of moving sheets
	Recorded information on ice behaviour and movement, especially ice jams	Seek evidence of logs and debris and their effect on flood waters
Geotechnical data	Excavation and pile-driving records from existing structures Drilling records from wells	Sample bed-material and photograph it in situ Seek evidence on largest size of stone moved by floods
	Soil test records	Describe and photograph bank materials Seek evidence of rock outcrops Arrange for subsoil investigations to maximum anticipated scour depths
Channel and coastal	Compare maps and airphotos of different years for evidence of channel shifting, movement of bars and spits, bank and shore erosion, etc.	Measure maximum scoured depths at bends, constrictions, cliffs, and existing structures Seek local evidence of channel shifting, bank and shore erosion, land slides, etc., and their causes Seek indications of general bed degradation or aggradation Observe nature and movement of bed forms
Engineering and control works (on regulated streams)	Operating procedures for hydraulic structures, both normal and emergency	Seek evidence of unrecorded engineering works, dredging, gravel mining, straightening, flow diversions, etc.
	Proposals for extended or new control works	Examine structural condition of minor impounding structures
Drainage basin and meteorological data (mainly applicable in regions with poor hydrometric records)	Drainage area above site Present land cover and land use, and anticipated changes therein Slopes, soil types, permeabilities Storage in lakes, reservoirs, etc. Possible future engineering works affecting run-off and flows Intensities and distribution of storm precipitation Snowfall, temperature, and other factors affecting snowmelt run-off Wind data affecting wave heights	Check land cover and use in field May require field checks

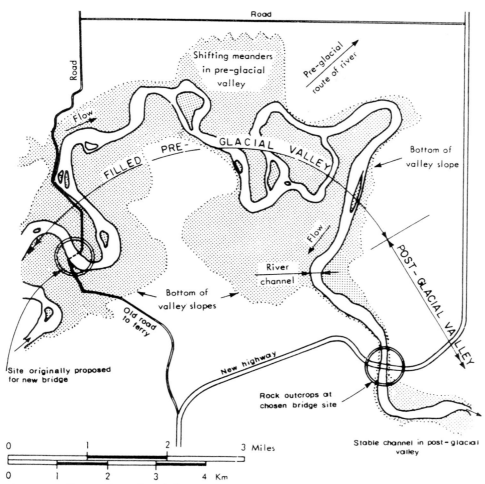

Figure 3.9. Changes in geology influencing the choice of location for a bridge (Neill 1975).

Neill also gives a review of basic design data for bridge designs, such as:

— bridge and road location in relation to stream channel characteristics (bends, alluvial forms), geology (Figure 3.9), the presence of existing bridge, plans for river works upstream or downstream such as dams, and sediment extraction.
— design high water level and bridge height, including effects of debris and ice.
— design discharge.
— length of bridges, obtaining an optimum solution between bridge construction cost and cost of river training works.
— back-water effects.
— navigation requirements.

For more details see Neill (1975).

REFERENCES

Ashmore, P. & G. Parker 1983. Confluence scour in coarse braided streams. *Water Resources Research* 19(2); 392–402.
Bendegom, L. van 1947. Enige beschouwingen over rivier morphologie en rivier verbetering. *De Ingenieur* 59(4); 1–11.
Callander, R.A. 1978. River meandering. *Ann. Rev. Fluid Mech.* 10; 129–158.
Cunge, J.A., F.M. Holly Jr. & A. Verwey 1980. *Practical aspects of computational river hydraulics.* Pitman, London.
Engelund, F. 1974. Flow and bed topography in river bends. *Proc. ASCE* 100(HY11); 1631–1648.
Garde, R.J., & K.G. Ranga Raju 1977. *Mechanics of sediment transportation and alluvial stream problems.* Wiley Eastern Ltd., New Delhi.
Gill, M.A. 1981. Bed Erosion in Rectangular Long Contraction. *Proc. ASCE* 107(HY3); 273–284.
Jansen, P.Ph. (ed.) 1979. *Principles of River Engineering.* Pitman, London.
Joglekar, D.V. 1971. *Manual on river behaviour, control and training.* Central Board of Irrigation and Power. Publ. no. 60. New Delhi.
Klaassen, G.J.R. Klomp & J.J. van der Zwaard 1982. Planning and environmental impact assessment of river engineering works via mathematical models. *Proc. 3rd Congress Asian Pacific Regional Div. IAHR, Bandung* (also Delft Hydr. Lab. Publ. 273).
Klaassen, G.J. & K. Vermeer 1988. Confluence scour in large braided rivers with fine bed material. *Int. Conf. on Fluvial Hydraulics, Budapest.*
Laursen, E.M. 1960. Scour at bridge crossings. *Proc. ASCE* 86(HY2); 39–54.
Leopold, L.B. et al. 1964. *Fluvial processes in geomorphology.* W.H. Freeman, San Francisco.
Neill, C.R. 1975. *Guide to bridge hydraulics.* Univ. of Toronto Press, Toronto.
Norman, V.W. 1975. *Scour at selected bridge sites in Alaska.* U.S. Geol. Survey, Water Res. Inst. 32–75.
Odgaard, A.J. 1981. Transverse slope in alluvial channel bends. *J. Hydr. Eng., ASCE* 107(HY12); 1677–1694.
Odgaard, A.J. 1982. Bed characteristics in alluvial channel bends. *J. Hydr. Eng., ASCE* 108(HY11); 1268–1281.
Odgaard, A.J. 1984. Flow and bed topography in alluvial channel bend. *J Hydr. Eng., ASCE* 110(HY4); 521–535.
Simons, D.B. & F. Sentürk 1977. *Sediment transport technology.* Water Resources Publications, Ft. Collins, Colorado.
Sharma, J.M. 1972. *Scour around bridge piers.* Indian Min. of Railways, Bridges and Floods Report No. RBF-10.
Shen, H.W. 1979. *Modelling of rivers.* Wiley, New York.
Schumm, S.A. 1977. *The fluvial system.* Wiley, New York.
Struiksma, N. et al. 1985. Bed deformation in alluvial channel bends. *J. Hydr. Res.* 23(1); 57–93.
Thomas, W.A. 1977. HEC-6. *Users manual: Scour and deposition in rivers and reservoirs.* The Hydrologic Engineering Center, Davis, California.
Vanoni, V.A. (ed.) 1975. *Sedimentation Engineering.* ASCE, New York.
Webby, M.G. 1984. General Scour at a Contraction. Bridge Design and Research Seminar, National Roads Boards. *N.A. RRU Bulletin* 73; 109–118.

CHAPTER 4

Scour around spur dikes and abutments

H.N.C. BREUSERS

4.1. INTRODUCTION

Structures such as groynes, dikes, guide banks, abutments and piers are often required in river banks in conjuntion with river training or road construction works. All such works concentrate the river flow and can therefore cause local scour. Scour around bridge piers, with a wealth of information available, is discussed in Chapter 5. In comparison, little information exists for scour in the vicinity of spur dikes and abutments. Furthermore, most of the information that exists was obtained from laboratory tests at greatly reduced scales; hence its extrapolation to full-scale conditions has an associated risk.

4.2. SPUR DIKES

Spur dikes are built out from the bank of a river to deflect the main river current away from an erodible bank. Another type of dike, called a guide bank, is built around a structure such as an abutment to protect it from erosion. Dikes, in general, constrict the flow in a river and increase both the local velocities near them and the mean velocity in the main stream.

Typical conditions for a spur dike and a guide bank are shown in Figure 4.1a and b for a channel with a rectangular cross section. In this simplified case, and if jet contraction is neglected,

$$Q = q_o B = q_1 (B-b)$$

Also the quantity

$$M = \frac{B-b}{B}$$

and the sediment diameter d are used in the following discussion. For practical purposes attention is focussed on those results which were obtained with sediment transport in the normal channel as well as in scour holes, i.e. on tests with $U > U_c$.

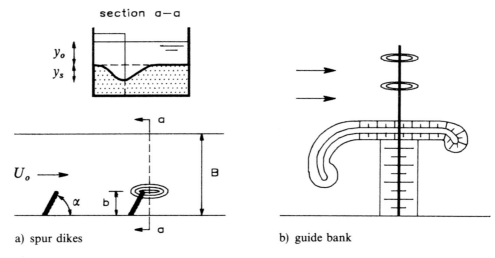

Figure 4.1. Definition sketch.

Inglis (1949) analysed field data on the maximum scour depth observed near spur dikes and guide banks in India and Pakistan. He compared the total scoured depth, $y_o + y_s$, with the three-dimensional Lacey regime depth, y_{3r}, which can be obtained from the equation

$$y_{3r} = 0.47 \left(\frac{Q}{f}\right)^{1/3} \tag{4.1}$$

In Equation 4.1, the silt factor, f, is equal to $1.76 \sqrt{d}$*

The ratio $(y_o + y_s)/y_{3r}$ ranged from 1.6 to 3.9, and Inglis recommended the use of the following values:

Scour at straight spur dikes angled upstream ($\alpha > 90°$) with steeply sloping noses (1.5V:1H)	3.8
Scour at similar dikes but with long sloping noses	2.25
Scour at guide bank noses of large-radius	2.75

Ratios over the observed range should be used with judgement as to the severity of the river's attack on the structure.

Laboratory studies of these structure were performed in flumes with fixed vertical side walls and erodible beds, and they could therefore be compared with Lacey's two-dimensional regime depth,

---

*If d is expressed in mm, Equation 4.1 serves for both metric and feet units.

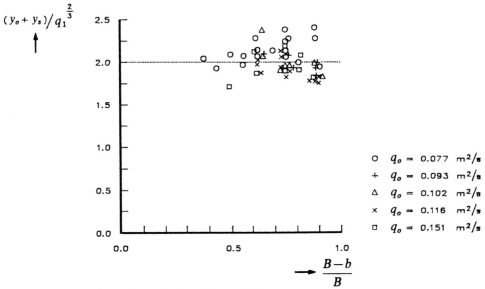

Figure 4.2. Experimental results for thin vertical spurs.

$$y_{2r} = 1.34 \left(\frac{q_1^2}{f}\right)^{1/3} \quad (4.2)$$

in which q_1 is the discharge per unit width in the contracted section, as noted.*

The most readily useful of the available studies are those of Ahmad (1953) which provided, particularly, information on the effect of the angle α on the depth of scour, and of Liu et al. (1961) whose results cover the widest range of the pertinent variables. Most of the available results are for spur dikes in the form of a vertical wall. Various other studies have added marginally to the limited information on this subject: Laursen and Toch (1956), Laursen (1958, 1963), Field (1971), Veiga da Cunha (1971), Karaki et al. (1974) and Richardson et al. (1975). In a related study Nwachukwu and Rajaratnam (1980) observed dramatic increases in boundary shear stress close to the end of a spur dike.

Ahmad presented his results on the basis of an equation with the form

$$y_o + y_s = K q_1^{2/3} \quad (4.3)$$

one which is compatible with the Lacey regime equation (K taking the place of $1.34/f^{1/3}$). His results for the effect of angle are summarized, with other relevant factors, at the end of this section.

Liu et al. (1961) performed extensive investigations on thin vertical spurs and other

---

* Coefficient is for the metric system. For units based on feet it should be multiplied by 2/3.

abutment shapes. Two flumes were used, with $B = 4.2$ m and 2.4 m, and two sands with average diameters, d_{50}, of 0.56 mm and 0.65 mm. Many of the tests were conducted using spur dikes formed from vertical boards (thus with a negligible thickness); the velocities were greater than U_c, throughout these tests. Although they presented their results in the form of a dimensionless empirical equation, utility remains in the Lacey type equation since y_o is not readily determinable in the field. A replot of their data in Figure 4.2 with K as a function of M indicates (a) that most of the data fall in the rather narrow range of $K = 2.0 \pm 10\%$ and (b) that no observable trend with other parameters occurs.

Other factors affect the maximum depth of scour. Ahmad considered the effect of asymmetry in the velocity distribution such as could occur downstream from a bend. He also tested different shapes of the end of the spur (T-shape, hockey spur) and found their effect to be less than 10%. He found no effect on the scour depth for the two grain sizes $d = 0.35$ mm and 0.70 mm. Liu et al. included abutments with sloping faces ($1V:1.5H$) and found that the resultant scour depths were only 50–60% of those for the vertical spurs. They also observed cases in which dunes formed and moved through the constriction. The values of y_s then fluctuated around an average about the same as that without dunes; the maximum values were greater than y_s by approximately $0.3y_o$.

Recommendation for design

Satisfactory results can be obtained from empirically determined values of K in the equation

$$y_o + y_s = K q_1^{2/3} \tag{4.4}$$

with a value of $2.0 \pm 15\%$ for a nearly vertical spur dike. Correction factors modify this result for various other conditions as follows:

Spur dike angle α

α	30°	45°	60°	90°	120°	150°
K_1	0.8	0.9	0.95	1.0	1.05	1.1

Shape of dike

	K_2
Vertical board	1.0
Narrow vertical wall	1.0
Wall with 45° side slopes	0.85

Position of dike	K_3
Straight channel	1.0
Concave side of bend	1.1
Convex side of bend	0.8
Downstream part of bend, concave side	
Sharp bend	1.4
Moderate bend	1.1

To an acceptable approximation, the combined use of the various factors (K, K_1, K_2, K_3) is recommended.

4.3. ABUTMENTS

The scour at abutments is similar in character to that around spur dikes. Most of the limited amount of information about abutments comes from studies that have been performed at the University of Auckland (Wong 1982, Tey 1984, Kwan 1984). Melville and Raudkivi (1984) presented a summary of these.

Kwan investigated the similarity between scour at a pier and at an abutment by comparing measured scour depths at a cylindrical pier and at a semi-circular abutment of the same diameter. The latter was attached to a short flat plate oriented parallel to the flow so as to simulate a wall. Scour depths with a short plate were slightly lower than for a circular pier, possibly due to the thin boundary layer that developed along the flat plate.

At bridge abutments along a continuous wall, the flow retardation adjacent to the wall weakens the downflow that occurs at the axis of symmetry for an isolated pier. The downflow and lateral flow components at the upstream face of the abutment combine to produce a strong spiralling vortex motion, shown schematically in Figure 4.3.

Most experimental data were obtained for $u_*/u_{*c}=0.9$ to 0.95 and $d=0.8$ mm (without general bed load). They showed that both the slope of the abutment and the ratio of water depth to abutment length are important, Figure 4.4.

For small values of y_o/b, the ratio of water depth y_o to the protrusion of the abutment into the river b, the depth of scour y_s increases almost linearly with y_o, whereas for y_o/b above about 1.0, y_s/b is nearly constant so that y_s is more nearly proportional to b.

Line 1 is fitted to data for abutments characterized as wing wall (WW), spill-through (ST) and vertical wall pointing downstream (TS1). Line 2 is for abutments which are semi-circular (SCE) and for vertical plates mounted perpendicular to the flow (TS2). The various abutments shapes are shown in Figure 4.5. The values of scour for the shapes of line 1 are about 70% of those for line 2 (semi-circular). The effect of particle size is small if $y_o/d>50$. If sediment is not uniform, the larger grains will have an armouring effect and reduce the scour. Quantitative results are mainly available for $u_*/u_{*c}=0.95$ (Wang 1982), which would give too large an effect if applied to larger u_*-values.

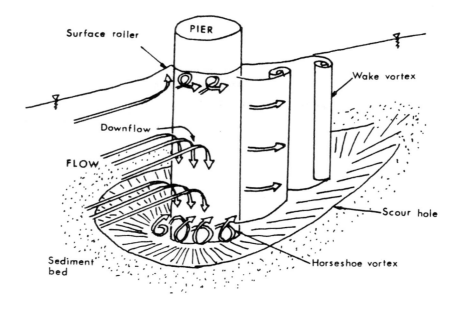
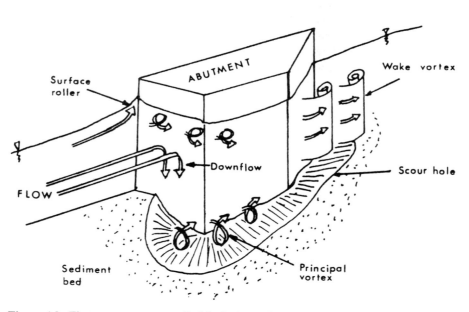

Figure 4.3. Flow structure at a cylindrical pier and at a wing-wall abutment.

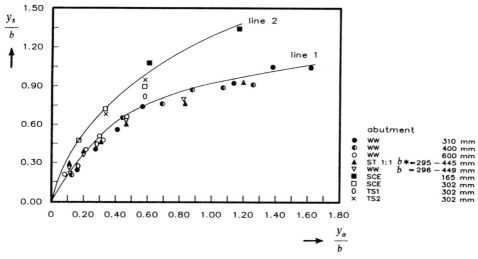

Figure 4.4. Effect of y_o/b on y_s/b for various abutments.

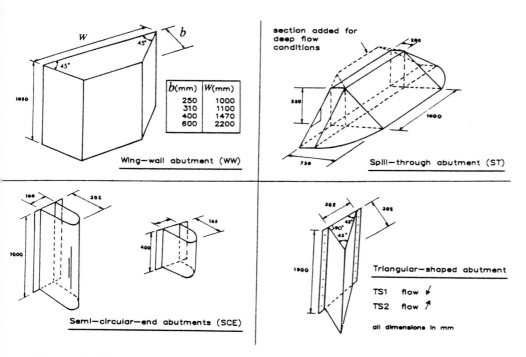

Figure 4.5. Abutment shapes.

Data from a current research project (Kandasamy 1989) indicate that the live-bed scour depth at a wing-wall type abutment varies with u_*/u_{*c}, where u_{*c} is the limiting critical shear velocity for clear-water scour conditions. At $u_*/u_{*c} = 1$ the scour depth has a maximum value. The scour depth decreases first when the $u_*/u_{*c} = 1$ value is exceeded and then increases again, analogous to the behaviour of scour at bridge piers. Some of the data are shown below:

Abutment length mm	Flow depth mm	$\dfrac{u_*}{u_{*c}}$	Scour depth d_s mm	$\dfrac{d_s}{d_{sc}}$
790	150	1.0	395	1
790	150	2.9	296	0.75
790	150	3.5	334	0.85
790	100	1.0	330	1
790	100	2.4	253	0.77
790	100	1.25	283	0.86
790	100	4.0	285	0.86
790	100	3.2	289	0.88
790	100	2.8	260	0.79
1200	100	1.0	440	1
1200	100	1.25	318	0.72

Recommendation for design of abutments

With so little information available, only limited recommendations can be made. The laboratory tests at Auckland show clearly that comparatively large scour depths can occur; these can be 2 to 5 times the depth of flow or 1 to 1.5 times the abutment length, depending upon its shape. Since the shape of the river bank and the abutment can vary widely, guidance from model tests may be necessary. Scour can be reduced by streamlining the abutment or using guide banks with a sufficiently large radius of curvature, so that flow separation or a strong contraction is avoided. If model tests are used, the similitude should be based not only on geometrical factors, but also on providing the same value of u_*/u_{*c} in model and prototype.

Abutments can also fail by a process known as outflanking, the erosion of the causeway behind the abutment, while the abutment remains intact. In this type of failure, the causeway material slides into the scour hole and creates a hole in the embankments behind the abutment. More of the flow can impinge on the eroded region, the strength of the downward flow increases and the scour hole increases in depth and width. The end result can be a breakthrough cutting off the abutment from the approach fill. If such a danger is envisaged, the causeway has to be protected with riprap or other types of revetments, or upstream spur dikes have to be provided to direct the river away from the causeway. Design examples from Indian rivers can be found in Joglekar (1971).

REFERENCES

Ahmad, M. 1953. Experiments on design and behaviour of spur dikes. *Proc. IAHR Conf. Minnesota;* 145–159.
Field, W.G. 1969. Flood protection at artificial river constructions. *ASCE. Nat. meeting on Transp. Eng., Washington,* paper 906.
Inglis, C.C. 1949. *The behaviour and control of rivers and canals.* C.W.I.N.R.S. Poona, Res. Publ. no. 13.
Joglekar, D.V. 1971. *Manual on river behaviour, control and training.* Central Board of Irrigation and Power. Publ. No. 60., New Delhi.
Kandasamy, J.K. 1989. *Abutment scour.* University of Auckland, Dept. of Civil Eng., Rep. No. 458.
Karaki, S. et al. 1974. *Highways in the river environment. Hydraulic and environmental design considerations.* Civil Eng. Dept. Colorado State Univ. Ft. Collins.
Kwan, T.F. 1984. *Study of abutment scour.* University of Auckland, Dep. of Civil Eng., Rep. No. 328.
Laursen, E.M. 1958. *Scour at bridge crossings.* Iowa Highway Res. Board, Bull. no. 8.
Laursen, E.M. 1963. An analysis of bridge relief scour. *Proc. ASCE* 89(HY3); 93–117.
Laursen, E.M. & A. Toch 1956. *Scour around bridge piers and abutments.* Iowa Highway Res. Board, Bull. No. 4.
Liu, H.K., F.M. Chang & M.M. Skinner 1961. *Effect of bridge construction on scour and backwater.* Colorado State Univ., Civil Engineering Section, Ft. Collins, CER 60 HKL 22.
Melville, B. & A.J. Raudkivi 1984. *Local scour at bridge abutments.* Nat. Roads Board New Zealand, RRU Seminar on bridge design and research, Auckland.
Nwachukwu, B.A. & W. Rajaratnam 1980. *Flow and erosion near groyne-like structures.* Dept. of Civil Eng., Univ. of Alberta, Edmonton, Alberta.
Richardson, E.V., M.A. Stevens & D.B. Simons 1975. The design of spurs for river training. *Proc. 16th IAHR Congress, Sao Paulo,* 2; 382–388.
Simons, D.B. & F. Senturk 1977. *Sediment transport technology.* Water Res. Publ., Ft. Collins, Colorado.
Tey, C.B. 1984. *Local scour at bridge abutments.* University of Auckland, Dept. of Civil Eng., Rep. No. 329.
Veiga da Cunha, L. 1971. *Erosoes localizades junto de obstaculos salientes de margens.* Ph.D. Thesis, Lisboa.
Wong, H.H. 1982. *Scour at bridge abutments.* University of Auckland, Dept. of Civil Eng., Rep. No. 275.

CHAPTER 5

Scour at bridge piers

A.J. RAUDKIVI

5.1. INTRODUCTION

The local scour at bridge piers has to be added to general scour and constriction scour to obtain the maximum scour depth for use in the design of bridge piers. In an analysis of local scour one must differentiate between clear-water scour and live-bed scour (as defined in section 1.2) because both the development of the scour hole with time and the relationship between scour depth and approach flow velocity depend upon which type of scour is occurring. Figure 5.1 shows the traditional diagrams by Chabert and Engeldinger (1956) for scour development with time and as a function of shear velocity or velocity. Raudkivi (1981) suggested that a second peak scour depth may exist as shown by the dotted line in Figure 5.1b.

Figure 5.2 is a diagrammatic presentation of laboratory test data on scour at cylindrical piers in *uniform sediments* from the University of Auckland, which clearly show the second peak. The results for grain sizes which lead to the formation of ripples ($d < \sim 0.7$ mm) differ from those for larger grain sizes for which ripples do not form. With non-ripple-forming sediments, experiments can be run with $u_* \cong 0.95 u_{*c}$ without the upstream bed being disturbed by the approach flow. With finer sands a flat bed cannot be maintained for these conditions. Ripples will develop on the bed of the approach flow, and a small amount of sediment transport will take place. Usually ripples develop at shear velocities u_* above $0.6 u_{*c}$. Thus, the clear water conditions are not maintained long enough for sands to reach the same peak scour depth that occurs for the coarser non-ripple-forming sediments. Clear-water experiments have to run continuously for several days before equilibrium conditions are approached. An exception (as discussed later) occurs if the geometric standard deviation of the sand sizes $\sigma_g \cong 1.3 - 1.5$. In this range, the coarser grains armour the surface but are not large enough to armour the scour hole where the agitation is higher. Then, clear water scour depths of the same order as observed with non-ripple-forming sediments can be reached in laboratory tests. The second peak occurs at transition flat-bed conditions and does not exceed the first except for fine sands, i.e. the second peak does not exceed about 2.3 times the diameter or width of the pier. The band, indicated by dotted lines signifies the range of fluctuations in scour depth due to the passage of bed features. The

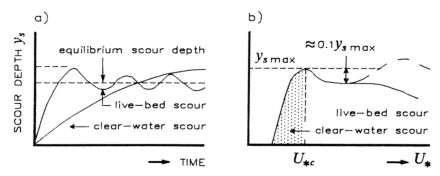

Figure 5.1. Scour depth for a given pier and sediment size, (a) as a function of time and (b) as a function of shear velocity or approach velocity.

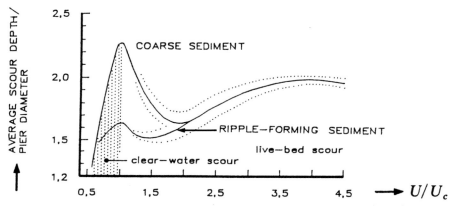

Figure 5.2. Diagrammatic illustration of scour depth at a cylindrical bridge pier in a uniform sediment.

fluctuations in depth at the transition flat bed do not go to zero because of avalanching of sediment in the scour hole.

In general, for the live-bed case, the scour increases rapidly with time (Figure 5.1a) and then fluctuates about a mean value in response to the passages of bed features; the equilibrium scour depends on the size of bed features, i.e. on variations in the depth of flow and is approximately y_{se} plus half the height of bed features.

5.2. FLOW PATTERNS AT A CYLINDRICAL PIER

The flow pattern past a pier protruding from a plane boundary in uniform open channel flow is complex. This complexity increases with the development of the scour hole. A detailed study of these flow patterns was carried out by Melville in order to gain a better understanding of the flow mechanisms which lead to formation of the scour hole. Results of the studies of flow pattern at a cylindrical pier have been reported by Hjorth (1972, 1975), Melville (1975), and Melville and Raudkivi (1977). Additional work was done by Ettema (1980). In order to aid discussion, the flow pattern is separated into its components:

(1) Downflow in front of pier
(2) Horseshoe vortex
(3) Cast-off vortices and wake
(4) Bow wave

These are shown in diagrammatic form in Figure 5.3.

In the vertical plane of symmetry in the direction of flow, the flow comes to rest at the face of the pier. Since the velocity, u, decreases from the surface downwards, the stagnation pressures (equal to $\rho u^2/2$) also decrease from the surface downwards, i.e. a downward pressure gradient. The strength of the downflow in front of the cylinder reaches a maximum just below the bed level when a scour hole is present. The maximum velocity of the downflow at any elevation occurs at 0.05 to 0.02 pier diameters upstream of it, being closer to the pier lower down. Its maximum value reaches 0.8 times the mean approach flow velocity and occurs in the scour hole at about one pier diameter below the bed level.

The so-called horseshoe vortex develops as the result of separation of flow at the

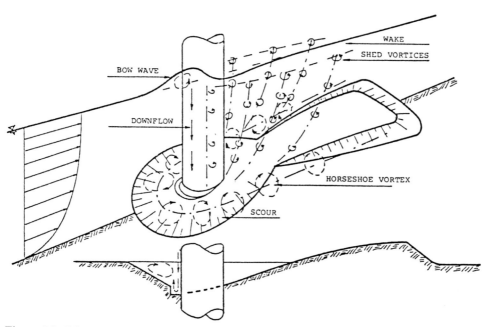

Figure 5.3. Diagrammatic representation of the flow pattern adjacent to a cylindrical pier.

upstream rim of the scour hole excavated by the downflow. It is a lee eddy similar to the eddy or ground roller downstream of a dune crest. The horseshoe vortex is a consequence of scour, not the cause of it, although it becomes effective in transporting material away from the scour hole. The horseshoe vortex extends downstream, past of the sides of the pier, for a few pier diameters before losing its identity and becoming part of general turbulence. In the scour hole, the horseshoe vortex pushes the maximum downflow velocity still closer to the pier.

The flow separates at the sides of the pier and the separation surfaces enclose the wake downstream of the pier. The separation creates a surface with a discontinuity in the velocity profile and this leads to the development of the concentrated "cast-off" vortices in the interface between the flow and the wake. The translation by the mean flow bends the axes of the vortices. Near the bed, these vortices interact with the "horseshoe vortex" causing the trailing parts to oscillate laterally and vertically at the frequency of vortex shedding, n, as given by the Strouhal relationship $nb/U \cong 0.2$. The cast-off vortices with their vertical low pressure centres lift sediment from the bed like miniature tornados.

Scour hole development commences at the sides of the cylinder with the holes rapidly propagating upstream around the perimeter of the cylinder to meet on the centreline. The eroded material is transported downstream by the flow. Soon after the commencement of scouring, a shallow hole, concentric with the cylinder, is formed around most of the perimeter of the cylinder (about $\pm 120°$) but not in the wake region. The downflow acts like a vertical jet eroding a groove in front of the pier, Figure 5.3. The eroded material is carried around the pier by the combined action of accelerating flow and the spiral motion of the "horseshoe vortex". The downflow is turned 180° in the groove and the upward flow is deflected by the "horseshoe vortex" in the upstream direction, up the slope of the scour hole. At this turning point, the lip of the groove is often very sharp and the face is almost vertical. The groove becomes shallow or disappears altogether when scour approaches its equilibrium depth. The rim collapses irregularly in local avalanches of bed material. The deflection of the downflow ejects this material up to where the "horseshoe vortex" tends to push some of it up the slope. The rest is picked up by the flow which carries it into and behind the wake region where a bar develops. The upstream part of the scour hole develops rapidly and has the shape of a frustrum of an inverted cone with slope equal to the angle of repose of the bed material under erosion conditions.

For clear-water scour, equilibrium is reached when the combined effect of the temporal mean shear stress, the weight component and the turbulent agitation are in equilibrium everywhere.

For live-bed scour, an excess shear stress must exist to transport the sediment through the scour hole. Particles on the surface of an equilibrium clear-water scour hole may be occasionally moved but are not carried away.

5.3. SCOUR FORMULAE FOR STEADY FLOW

The scour depth depends on variables which characterize the fluid, bed material, flow and bridge pier, e.g.

$$y_s = f(\rho, v, g, d, \rho_s, y_o, U, b) \tag{5.1}$$

or, in dimensionless form:

$$\frac{y_s}{b} = f\left(\frac{Ub}{v}, \frac{U^2}{gb}, \frac{y_o}{b}, \frac{d}{b}, \Delta\right)$$

$$= f\left(\frac{u_* b}{v}, \frac{u_*^2}{g\Delta d}, \frac{y_o}{b}, \frac{d}{b}, \Delta\right) \tag{5.2}$$

where $u_* = (gy_o S)^{1/2}$ is the shear velocity and other terms are defined in the notation. Literature surveys are published by Breusers et al. (1977) and Raudkivi and Sutherland (1981). The published formulae for scour at bridge piers are based on various combinations of such parameters as appear in Equation 5.2.

Some formulae are for *clear-water scour*, some for *live-bed scour*, and some are intended to serve for both. Unfortunately, the formulae offer little guidance to the designer. The scour depth predicted for given conditions can differ by a factor of five or more, as has been demonstrated, for example, by Melville (1974), Anderson (1974) and Hopkins et al. (1975, 1980). This diversity is probably the reason that Neill (1973) defined the local scour depth at bridge piers simply as follows:

y_s/b	Pier shape in plan
1.5	Oblong with rounded nose
1.5	Cylindrical
2.0	Rectangular with square nose
1.2	Ogival (sharp nose)

These values for the ratio of local scour depth y_s to the width of the pier, b, are for flow aligned with the pier. The effects of a flow approaching the pier at an angle will be discussed later as one of the effects due to individual parameters. Since the local scour is the result of very complex interactions of a great many parameters, one must consider the effects of each parameter. An understanding of these will enable the designer to make refinements and thus minimize the amount of scour depth to be expected in specific circumstances.

5.4. EFFECT OF SPECIFIC PARAMETERS

The large range of predicted and measured values of local scour depths at bridge piers, for what appear to be identical conditions, can be explained by the effects of a variety of specific parameters. These include characteristics of the sediment, the flow and the geometry of the pier and the channel.

5.4.1 Effect of sediment grading

The effect of grain size distribution on the local depth of scour at a bridge pier was studied by Raudkivi and Ettema (1977a, b). In their tests, the pier diameter was 102 mm and the flume was 1.5 m wide. The experiments were conducted so that the shear velocity was nearly equal to but less than that which would cause general transport, i.e. for clear-water conditions.

Figure 5.4 summarizes the results obtained for the maximum clear-water scour depth, relative to the pier diameter y_s/b, as a function of the sediment characteristics, σ/d_{50} in which σ is the standard deviation of the grain-size distribution and d_{50} is the median particle size.

The maximum depth of scour for coarse grained, non-ripple-forming sediment ($d > 0.7$ mm) of a single grain size ($\sigma \to 0$) is $y_s/b \cong 2.1\text{--}2.3$ and does not depend on the grain size. If the sediment is a uniform sand ($\sigma \to 0$) with grain size $d < 0.7$ mm, a flat bed cannot be maintained near threshold shear stress, ripples develop, and a small amount of general sediment transport takes place and replenishes some of the sand scoured at the pier. Thus, true clear-water scour conditions cannot be maintained for this case. The resultant equilibrium scour depths were $y_s/b \cong 1.4\text{--}1.5$.

The use of an upstream shear stress which is the maximum for which no ripples develop reduces scouring power of the flow and reduces the maximum scour depth to about the same value as that for flow with limited transport.

As the standard deviation of the grain size distribution increases, the larger sand grains form, in due course, an armour layer on the upstream bed and prevent the

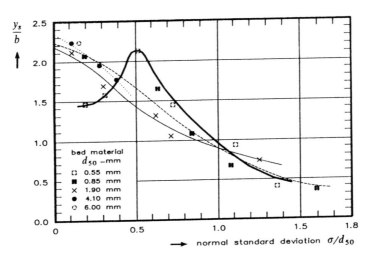

Figure 5.4. Equilibrium clear-water scour depth divided by pier diameter (y_{se}/b) as a function of the sediment grading, $u_* \cong u_{*c}$.

development of ripples. There is a critical value of σ/d_{50} for which armouring can just be achieved on the plane bed but not in the scour hole where the local forces on the grains are higher due to increased turbulent agitation.

For $d_{50} = 0.55$ mm sand, the critical conditions were achieved when $\sigma/d_{50} \cong 0.5$. For this case the maximum scour depth was $y_s/b \cong 2.1$ after 160 hr. For larger values of σ/d_{50} the equilibrium clear-water scour depth, y_s/b, dropped to values in the range 0.3 to 0.5 for all sediments. The maximum clear-water scour depth shown in Figure 5.4 is larger than values reported in the literature. For example, Breusers (1965) gives, for live-bed scour in fine sand at cylindrical piers, $y_{se}/b = 1.4$. This value is correct for a particular value of σ/d_{50}. However, clear-water scour depths can have any value in the range of $0.3 < y_s/b < 2.3$. The probable scour depth can be reliably estimated only if the value of σ/d_{50} of the bed material is known.

Clear water scour develops very slowly and displays various stages as discussed by Raudkivi and Ettema (1977a, b). Grading affects the maximum value of y_s/b but not the time taken to reach equilibrium.

In practice the possible maximum value of the equilibrium depth of clear-water scour, $y_{se}(\sigma)/b$, can be estimated from

$$\frac{y_{se}(\sigma)}{b} = K_\sigma \frac{y_{se}}{b} \qquad (5.3)$$

where y_{se} is the equilibrium scour depth in uniform sediment, $\sigma_g \cong 1.0$. The K value as a function of σ_g depends on whether the sediment is ripple forming or not. The two values of K_σ are shown in Figure 5.5 for sediments which do or do not form ripples. Figure 5.5 is a re-plot of Figure 5.4 in terms of σ_g. Although the data were acquired with one pier size only ($b = 102$ mm), the trends are considered to be representative. When $u_*/u_{*c} < 0.8$, the formation of an armour layer one grain thick on the surface of non-uniform ripple-forming sands prevents the growth of ripples and the scour develops as in a non-ripple-forming sediment.

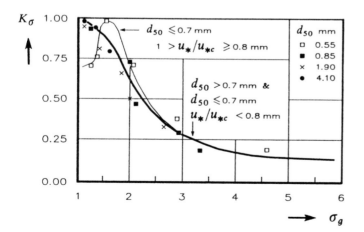

Figure 5.5. Coefficient K_σ as a function of the geometric standard deviation of the particle size distribution.

Under live bed conditions the major effects of sediment grading are the following:

(1) The grain size and grain size distribution affect the type and height of bed features and hence, the range of variation of scour depth from its mean value under given conditions;
(2) The grain size distribution affects the armouring process of the bed and hence, the mean scour depth at given conditions of flow;
(3) If the larger grains of the distribution are near the threshold condition, they tend to accumulate in the local scour hole. This increases the porosity of the bed and more of the downflow disperses there, i.e., its ability to scour is reduced;
(4) If the large grains become large relative to the pier size, the local scour depth decreases as discussed in Section 5.4.2.

5.4.2. Effect of pier and sediment sizes

The development and equilibrium depth of local scour are modified by the relative size, b/d_{50}, of the pier and sediment. Laboratory data (Ettema 1980), plotted for six pier sizes and sediment sizes from $d_{50} = 0.24$ mm to 7.80 mm in Figure 5.6, show that the maximum value of clear-water scour $(y_s/b)_{max}$ is unaffected by particle size as long as the value of b/d_{50} is larger than about 25. For smaller values, the grains are large compared to the groove excavated by the downflow, and the erosion process is impeded. The initial phase of scour develops similarly for most values of b/d_{50}, but the principal erosion phase and equilibrium phase are affected by the ratio of b/d_{50}. The experimental results for maximum clear-water scour, Figure 5.6, were subdivided into four groups and show a clear trend for non-ripple-forming sediments.

With ripple-forming sediment the scour depths are lower, as discussed before, and the results are more scattered. If $b/d_{50} < 8$ the stones are relatively so large that the scour is mainly due to the entrainment at the flanks of the pier. For coarse sediment the stones are again large relative to the width of the groove excavated by the downflow and the bed is fairly porous. A significant fraction of the downflow penetrates the coarse bed material and dissipates its energy. The peak occurs at about $b/d_{50} = 50$. For higher ratios Ettema showed a small reduction in the equilibrium scour depth but there is some doubt about this reduction; e.g. Chee (1982) recorded $y_{se}/b \cong 2.3$ for $b/d_{50} \cong 100$. Chiew (1984) carried out additional tests at live-bed conditions. His data concur with the above clear-water relationship, except that no peak was observed, the function just approaches a limiting value at about $b/d_{50} = 50$. When all data are normalized with the limiting scour depth values at $b/d_{50} = 50$, i.e. ignoring the decrease of clear-water scour depth at larger values as suggested by Ettema, all data collapse more or less into one function. The effects of ripples are apparent only near the threshold region. This normalized function is included in Figure 5.6 where $K(b/d_{50})$ is an adjustment factor for scour depth due to relative sediment size effect.

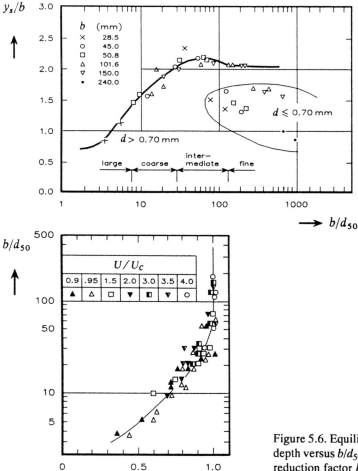

Figure 5.6. Equilibrium clear-water scour depth versus b/d_{50} at $u_*/u_{*c} = 0.90$ and the reduction factor K for scour depth due to b/d_{50} for clear water and live bed conditions.

The effect of pier size on the depth of local scour is of major interest when laboratory data are interpreted for field use. Pier size affects the time required for the local clear-water scour to reach the ultimate depth, not its relative magnitude y_s/b, if the effects of relative depth, y_o/b, and relative grain size b/d_{50} on the local scour depth are excluded. The actual value of y_{se}/b also depends on the grading of the bed material as discussed in Section 5.4.1. In the case of live-bed scour, the time taken to reach scour equilibrium is strongly dependent on the ratio of bed shear stress to critical shear stress and comparisons can be made only for the same values of this ratio.

The volume of the local scour hole formed around the up-stream half of the perimeter of the pier is proportional to the cube of the pier diameter (or the projected width of the pier). The larger the pier, the larger the scour volume and the longer is the time required for the scour to develop at a given shear stress ratio. Laboratory data ($b \cong 50, 100, 150$ and 240 mm) and uniform grain sizes show that the time to attain a certain depth, (y_s/b) is a function of b^3

$$\frac{y_s}{b} = A \ln[(d_{50}v/b^3)t] + B \tag{5.4}$$

where A and B are constants. Thus, for a given particle size and value of y_s/b, the value of $t \propto d^3$. Data plotted as (y_s/b) versus $(d_{50}v/b^3)t$ yield an almost single function.

Clear-water scour may take three to four days to reach equilibrium conditions under $u_*/u_{*c} \cong 0.90$ to 0.95. Live-bed scour develops very rapidly. In the laboratory, steady state conditions can be established in a few minutes at high transport rates of sediment, i.e. $u_* \gg u_{*c}$.

A special case arises when the bed material is highly suspendable, as for fine sand in a large river or polystyrene particles in a laboratory flume. Then the depth of the scour is likely to depend on the intensity of turbulence and the strength of eddying around the pier. The scouring is further aided by the "vacuum cleaner" effect of the cast-off vortices at the rear of the pier. These can lower the bed at the rear of the pier and thus ease the removal of sediment around the front half. The process could also lead to a deeper scour at the rear than in front, should the front part become armoured. However, the writer has not found any data on systematic studies of this aspect of the scour problem at bridge piers.

5.4.3. Effect of flow depth

The complex three-dimensional pattern of flow past the pier precludes an analytical determination of the effect of flow depth y_o on the maximum depth of local clear-water scour, y_{se}, except maybe by numerical simulation techniques. Observations show, nonetheless, that for shallow flows, the scour depth increases with depth of flow but, for larger water depths, the scour depth is almost independent of depth.

The influence of depth is assumed to depend predominantly on the ratios of u_*/u_{*c} and y_o/b. Breusers et al. (1977) discuss the influence of y_o/b but do not distinguish between clear-water and live-bed scour. For a constant u_*/u_{*c}, the influence of flow depth can be neglected for y_o/b greater than 2 to 3. For non-uniform sediments u_{*c} has to be replaced by the shear velocity which develops the coarsest armour layer on the bed.

Neill (1964) used the data by Laursen and Toch to show that the depth of local scour is a function of depth of flow, for *constant discharge*:

$$\frac{y_s}{y_o} = 1.5 \left(\frac{b}{y_o}\right)^{0.7} \tag{5.5}$$

This relationship, at least partly, arises from a transition from live-bed to clear-water scour. The argument used was that the approach velocity at the grain level decreases while the strength of the "horseshoe vortex" is affected little by the increase in flow depth and hence, the velocities at the grain level in the scour hole remain unchanged. The decreasing rate of sediment supply into the scour hole from upstream leads to an increased scour depth.

A pier causes a surface roller, like a bow wave to a boat, and a "horseshoe vortex" at the base of the pier. These two rollers have opposite senses of rotation. In principle, as long as the two rollers do not interfere, the local scour depth is insensitive to depth of flow. With decreasing depth of flow, the surface roller becomes the more dominant. The roller in the scour hole, the "horseshoe vortex" is still maintained but the bow wave interferes with the downflow which becomes weaker.

The stagnation pressure increases with the velocity of approach flow, and for a constant ratio of u_*/u_{*c}, with the particle size.

In a shallow flow, the equilibrium scour depth decreases with particle size for the same values of u_*/u_{*c} and y_o/b as illustrated by Figure 5.7. The finer the sediment relative to pier size, the smaller is the range of influence of flow depth. For fine sediments, the scour depth may be almost independent of flow depth when $y_o/b \cong 2$, whereas for relatively coarse sediments, this ratio may be closer to six. For very shallow flows, the formation of a bar downstream of the pier also causes a decrease in the depth of scour. Laboratory data available with live bed up to transition flat bed conditions indicate the same trends for depth dependence as shown in Figure 5.7. At transition flat bed conditions, the local scour depth at small values of y_o/b is about 10% less than for maximum clear water scour shown in Figure 5.7.

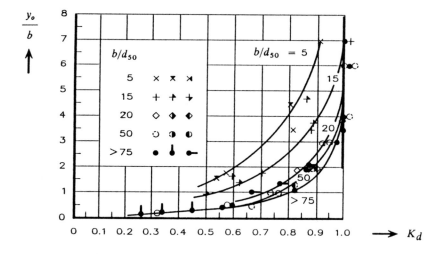

Figure 5.7. Equilibrium scour depth versus relative flow depth, y_o/b, with relative pier size b/d_{50} as parameter. 1) Ettema (1980), 2) Chiew (1984) and 3) Chee (1982).

5.4.4. Effect of pier alignment

The depth of local scour for all shapes of pier, except cylindrical, depends strongly on the alignment to flow. The depth of scour is a function of the projected width of the pier, i.e. the width normal to the flow. A long pier, misaligned with the flow, has an apparent width which is substantially greater than its actual thickness. As the angle of attack increases, the point of maximum scour depth moves along the exposed side of the pier towards the rear, and the scour depth at the rear becomes greater than that at the front. Figure 5.8. The angle of attack for which the scour becomes deeper at the rear, depends on the ratio of the length of the pier to its width.

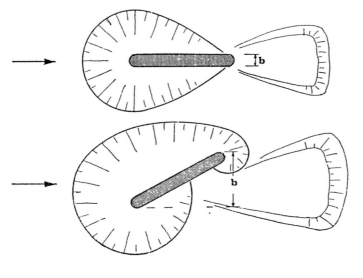

Figure 5.8. Diagrammatic scour shapes at a pier aligned with flow and another angled to the flow direction.

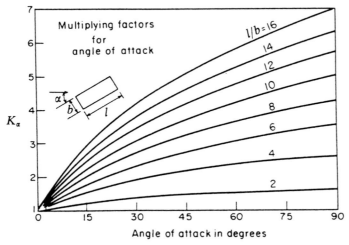

Figure 5.9. Alignment factor K_α for piers not aligned with flow (Laursen and Toch 1956).

The classic studies of the effect of shape, alignment and velocity profiles by Tison (1940) constitute a historic benchmark for numerous subsequent studies. Laursen and Toch (1956) published a graph of the multiplying factor for the scour depth K_a, as a function of the angle of attack, α, and the ratio of pier length to width, l/b, Figure 5.9.

In general, angles of attack greater than 5–10°, should be avoided. If this is not possible or if flow conditions can be assumed to change, the use of a row of cylindrical columns would be preferable to a solid pier (Section 5.8).

5.4.5. Effect of pier shape

Objects placed in water flow are often streamlined in order to reduce the drag exerted by flow and to reduce the size of the wake and general flow disturbance. Streamlining, however, is effective only as long as the body is aligned with flow within quite narrow limits.

The effect of pier shape, like that of alignment, can be accounted for by a shape factor K_s. The method is necessarily approximate because of the large variety of shapes, and especially because of changes in effective shape which can occur due to debris trapped during floods or to ice jams.

For piers aligned with flow, values of the shape factor K_s are summarized in Table 5.1 in terms of width to overall length ratio, b/l, and pier shape, Dietz (1972).

However, an appreciable range of the values of K_s has been reported.

Table 5.1. Pier shape coefficients.

Pier shape	b/l	b'/l'	K_s
Cylindrical			1.0
Rectangular	1:1		1.22
	1:3		1.08
	1:5		0.99
Rectangular with semi-circular nose	1:3		0.90
Semicircular nose with wegde-shaped tail	1:5		0.86
Rectangular with chamferred corners	1:4		1.01
Rectangular with wedge-shaped nose	1:3	1:2	0.76
		1:4	0.65
Elliptic	1:2		0.83
	1:3		0.80
	1:5		0.61
Lenticular	1:2		0.80
	1:3		0.70
Aerofoil	1:3.5		0.80

The cylinder is used for comparative purposes. In practice, shape factors are important only if axial flow can be assured. Even a small angle of attack of the approach flow will eliminate any benefit from pier shape.

The scour depth at the pier is also dependent on the slope of the leading edge of the pier to the vertical. Laboratory results for piers tapered as shown in Figure 5.10 are compared with a cylindrical pier of the same diameter as the thickness of the tapered piers ($b=45$ mm) in the same sand bed ($d_{50}=0.60$ mm, $\sigma_g=2.0$) and same flow conditions in Table 5.2. They show an increase in scour depth for an upwards broadening pier and a decrease for an upwards narrowing pier.

The process of local scour is frequently complicated by the presence of a footing, a caisson, or capping blocks on piles. The data on scour depths are predominantly based on well defined shapes, such as cylinders, rectangular piers, etc. without the three-dimensional effects that for example, a capping block may cause.

A footing or caisson with the top below the general bed level, and a diameter 1.6 times that of the pier or more, Figure 5.11, is effective in reducing the local scour depth by interception of the downflow. However, if the top of the caisson becomes exposed to the flow, the scour depth increases, i.e. the local scour depth is then essentially governed by the diameter or projected width of the caisson rather than of the pier.

The bed level to be considered is the level of the troughs of the largest bed features, a difficult quantity to determine for field conditions. The bed features vary a lot in size

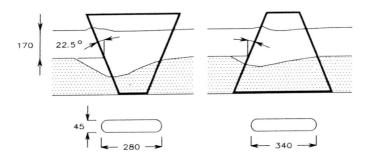

Figure 5.10. Scour at piers tapered in elevation. Dimensions at the original bed level.

Table 5.2. Shape factors for tapered piers.

Shape	Velocity m/s	y_{se}/b	K_s
Cylinder	0.67	1.69	1.0
	0.81	1.71	
▽	0.67	2.03	1.20
	0.81	2.06	
△	0.67	1.31	0.78
	0.81	1.26	0.74

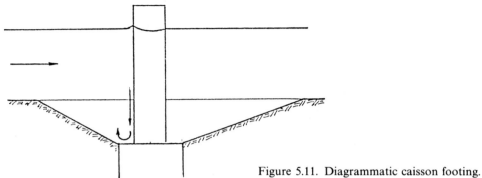

Figure 5.11. Diagrammatic caisson footing.

and shape, and three-dimensional patterns can lead to localized low spots. In addition, the level of the river bed can drop substantially during a flood in a particular stretch of the river, or the location of the pier may coincide with that of a moving stream channel in the cross section as discussed in Section 3.2. Thus reliance on such methods of scour reduction is dangerous indeed unless definite predictions of bed level are possible. In such cases the local scour depth should be based on the diameter of the caisson. If a caisson with a larger diameter is used, it should have a conical transition to the upper smaller diameter. The conical (or other) taper has an effect similar to that shown in Table 5.2.

Some writers have suggested that the depth of scour could be reduced by the use of a plate collar which could slide down and thus, protect the bed against scouring by the downflow. In practice however, such a device does not work. The downstream rim of the plate usually remains supported on the bar of the sediment deposited from the scour hole and water flows under the upstream edge of the plate. Hence little reduction in the scour depth is achieved.

5.5. LIVE-BED SCOUR

A distinction was made in the introduction between live-bed and clear-water scour but the preceding discussion relates mainly to the clear-water scour. Usually sediment transport occurs, both as bed and suspended load. Sediment from upstream flows into the scour hole and out again. The features of local scour at a pier for both clear-water and live-bed conditions are illustrated in Figure 5.2 in terms of the ratio of mean velocity of flow, U, to critical velocity, U_c, for initiation of sediment movement. The scour depth decreases from the threshold or *clear water peak* value at $U/U_c = 1$ to a minimum at about $U/U_c = 1.5 - 2$, at which stage the bed features are generally steepest; it increases again with further increase in the U/U_c-ratio to a new peak value at the transition flat bed condition, the *transition flat bed peak*. Over the transition flat bed the form drag component of energy loss is absent and a great fraction of flow energy goes into sediment transport and scouring. At still higher relative velocities (the upper flow regime), bed features appear again and dissipate some of the flow energy. The scour depth appears to decrease slightly but data on this phase of scouring is limited.

Under live bed conditions there are an average equilibrium scour depth, an average maximum scour depth and an average minimum equilibrium scour depth at each flow condition. The range of scour depths arises from the passage of bed features past the pier. Since the height of the bed features is a function of the flow depth only, and not of pier size, the normalized data for observed scour depths should be presented as y_s/b in terms of the *average equilibrium scour depth*. The range between the maximum and minimum values of local scour depths is approximately equal to the height of bed features under given flow conditons, i.e. the maximum depth is the sum of the average depth and half of the dune height.

Figure 5.12 shows experimental data by Chee (1982) for average maximum

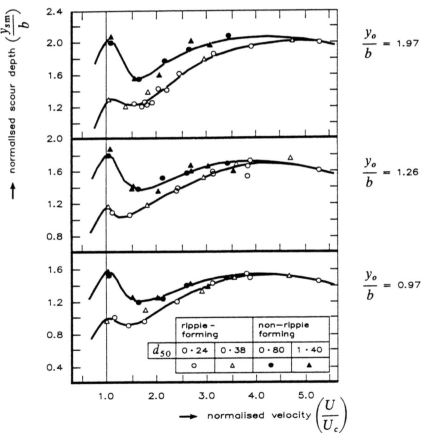

Figure 5.12. Laboratory data on average maximum local scour depth at a cylindrical pier.

Scour at bridge piers 77

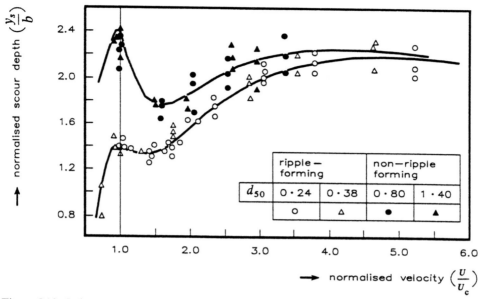

Figure 5.13. Laboratory data for average local scour depth at cylindrical piers in relatively deep water.

equilibrium depths of local scour obtained in the laboratory for uniform sediment at cylindrical piers (diameters were 50.8; 79.5 and 101.6 mm and the flow depth was 100 mm).

If the data are plotted as average scour depth and adjusted for flow depth effect, Figure 5.13 results. A practically identical plot was produced by Chiew (1984) for the local scour in uniform sediments using pier diameters of 32 and 45 mm in 170 mm depth of flow.

Figure 5.2 as well as Figures 5.12 and 5.13 are for subcritical flow conditions. In laboratory flumes where flow depth is limited the velocity may become super-critical (Fr > 1), particularly if the flume carries coarse sediment.

Both in the laboratory, and even more so in the field, great difficulties are encountered with the definition of the maximum local scour depth as distinct from fluctuations in bed level. In a laboratory flume, bed features can be observed to pass the pier, so that the mean bed level varies. The periods and wave lengths of such bed features are probably characteristic of a particular flume. Furthermore, the bed features sometimes assume a degree of three-dimensionality, and the flow in the flume meanders. This flow excavates deeper troughs at the side walls, spaced alternately from side to side. In flumes where the ratio of flume width to pier diameter is less than about fifteen, the local bed level can drop significantly, and the rapidly developing local scour superimposed on this bed level, may give extreme scour values; however these values are not transferable to field conditions.

The problems of definition are no less in the field. The lengths and heights of bed features can only be defined in statistical terms and oblique dunes and other three-dimensional features are the rule rather than the exception. Particularly large variations exist in the pool and bar profiles of rivers, in both depths and wave lengths. The bottom elevation of the local scour in the trough of a large bed or profile feature may indicate a large scour depth, whereas relative to the bed level, the scour depth may be small. Under high bed shear stresses, the local scour develops rapidly and the time of passage even of a large trough of a dune would be long enough for the local scour to stabilize. The picture can be more complicated in a river where local cut and fill processes occur during a flood and where shifting "stream channels" in the cross section occur as discussed in Section 3.2. Such moving channels are a feature of gravel rivers. For example, measurements on the Ohau River, in New Zealand, showed local channel depths up to 3 m in a flow of about 0.75 m depth and local scour depths up to 1.28 m relative to local bed level. The water surface elevation varied less than 0.5 metre. The writer knows of no methods by which the depth of such channels or the velocities in them could be reliably predicted.

The *effect of sediment grading* on the depth of local scour is still a subject of studies. An oversimplified conclusion is that as long as all grains are in motion, the equilibrium scour depth at the second transition flat bed peak is little affected. A proviso is that the large grains in transport must be smaller than about 0.1 of the pier diameter and the flow remains sub-critical. The grain-size criterion could be important in laboratory studies but is rarely a factor in the field. If the flow is supercritical, due to small flow depth at smaller velocity ratio U/U_c than that for the second peak in sub-critical flow, the scour depth at the second peak appears to have the value it would have had at the given U/U_c in sub-critical flow. The effect of sediment grading on the depth of live-bed scour is displayed in Figure 5.14. Generally, the effect appears to be negligible if $\sigma_g < 2$ and the scour follows approximately the uniform sediment relationship. For greater variations in sediment size, the scour at threshold, $U/U_c \cong 1$, is given by Figure 5.5 with U_c based on d_{50}-size. The d_{50}-size of a nonuniform sediment is only a rough indicator; selective sediment transport commences at lower values of u_{*c} than indicated by the d_{50}-size and some grains will resist substantially higher shear stresses. The commencement of sediment transport over a bed composed of nonuniform sediment also marks the beginning of the armouring process of the bed surface. The armouring increases the effective critical shear velocity that the bed surface can resist and the local scour depth increases up to a limiting or critical value of the shear velocity u_{*a}, or velocity U_a, of the armour layer for the given sediment. At this shear stress the local scour in nonuniform sediment reaches the first peak, the *armour peak* as in Figure 5.14.

At shear velocities $u_{*c} < u_* \leq u_{*a}$ the sediment transport can be *unsteady*, approaching zero as the armour layer stabilizes, or have a *dynamic equilibrium* state. In the latter condition the sediment that has been eroded from the surface remains in transport as, for example, in a sediment recirculating flume. In such a flume a fast moving layer of finer fractions can be seen to "slide" over the slowly evolving armour bed. This layer can even show bed features if the sediment is rich in finer fractions.

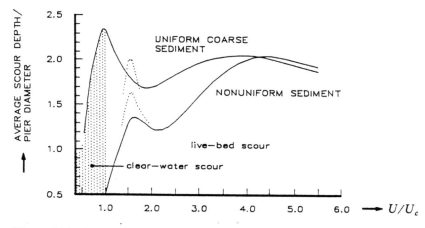

Figure 5.14. Average local equilibrium scour at a 45 mm diameter pier in uniform sediment ($\sigma_g \cong 1.3$) and in a nonuniform sediment ($\sigma_g = 3.5$), $d_{50} = 0.8$ mm, flow depth $y_o = 170$ mm. Full lines in both cases are for recirculating sediment transport. Dotted lines shows the armour peak with diminishing sediment transport.

The unsteady state of the armouring, for $u_* \cong 0.95 u_{*a}$, leads to similar scouring conditions and scour depths for the first scour peak as for clear-water scour peak if flow conditions last long enough for the limiting equilibrium condition to be established. The dynamic equilibrium condition leads to a lower first peak because of the amount of sediment in transport. The actual value of y_s/b depends on the grading of sediment. If the amount of sediment in transport is large, there could be a transport of finer sediment over a layer of coarser grains which are exposed only in the troughs of the bed features. However, when this layer breaks down, the rapidly increasing sediment transport rate leads to an initial decrease in the local scour depth.

On passing the critical shear stress, τ_{ca}, the surface layer is eroded and the sediment transport rate increases rapidly, leading to a reduction of local scour depth, analogous to transition from clear-water scour to live-bed scour.

After the initial decrease, the scour depth increases again with increasing applied shear stress up to the transition flat bed condition. First, the coarser fractions of sediment are still effective in armouring the bed of the local scour hole, but then their effectiveness decreases with increasing bed shear stress and vanishes at the transition flat bed conditions altogether; these trends are observed if the largest particles in the sediment are not large relative to the width of the groove excavated by the downflow (not larger than about one-tenth of the pier diameter).

Limited laboratory data indicate that the effect of sediment grading is too small to be distinguished from the usual scatter if $\sigma_g < 2$ and the data follow the uniform sediment function. On the other hand, field data indicate that fluid shear stresses in gravel rivers during floods seldom exceed about double that of the critical armour layer shear stress. Thus, if prolonged periods of flow at just below τ_{ca} are not likely, the actual scour depths observed may be appreciably less than the peak values according to Figure 5.14.

The critical velocity over a limiting armour layer, U_{ca}, can be estimated from d_{50}-size of this armour layer, d_{50a}. According to the studies by Chin (1985) the maximum value of d_{50a} is

$$d_{50a} = d_{max}/1.8 \tag{5.6}$$

where d_{max} is the characteristic maximum size of the bed material and has to be estimated. A method of doing this is to extrapolate the bed material grading curve on the basis of the last two or three data points to the 100% passing size. If the sieves used are in the $2^{1/4}$-series the error involved by assuming d_{max} to be equal to the coarsest sieve will be less than 20%. The limiting armour layer d_{50a}-size yields from the Shields criterion the u_{*ca}-value, and the corresponding critical mean velocity could be estimated from

$$U_{ca} = u_{*ca} \left[5.75 \log(y_o/2d_{50}) + 6 \right] \tag{5.7}$$

If the live-bed scour depth data for nonuniform sediment are normalized with u_{*ca} or U_{ca} the plot will look like that for uniform coarse sediment, except that for $U > U_{ca}$ a family of curves is obtained with the geometric standard deviation σ_g of the bed sediment size distribution as parameter. However, the results depend strongly on the ratios b/d_{max} and b/d_{50a} as well as on the grading of bed material, particularly on how porous the bed is.

The critical velocity at which the armour peak occurs, if some sediment is in transport over the armour layer, was found by Baker (1986) to be about $0.8\, U_{ca}$.

5.6. LOCAL SCOUR IN LAYERED SEDIMENTS

Many sedimentary deposits are non-homogeneous and, frequently, distinct layers are present. In some instances, coarse sediments cover deposits of fine grained sediment. It is of great practical interest to be able to predict what happens when the local scour at the bridge pier penetrates the layer of coarse sediment but very little in field or laboratory data is available on this aspect. Ettema (1980) studied this problem in the laboratory under conditions where the layer of coarse sediment was at the point of movement, i.e. clear-water scour, and separated the problem into four major segments, as shown in Figure 5.15:

(1) The covering layer is thicker than the local scour depth, i.e. the usual scour problem.
(2) The local scour penetrates the covering layer and triggers a disintegration of the layer in both upstream and downstream directions for a considerable distance. The end conditions are those of local scour in the underlying sediment due to the new hydraulic conditions.

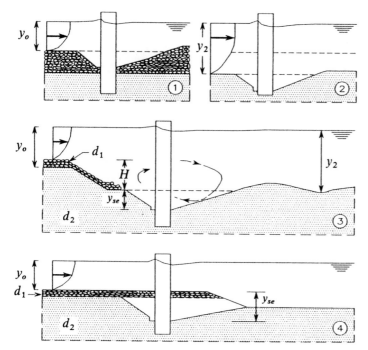

Figure 5.15. Diagrammatic forms of local scour in a layered bed.

(3) The covering layer disintegrates in downsteam direction only, leaving a step just upstream of the pier. The local pier scour develops at the bottom of this step.
(4) The covering layer disintegrates only over a small area downsteam of the pier but remains intact at both sides and upsteam.

The third case was found to give the deepest local scour relative to the initial bed level.

For the fourth case, the equilibrium scour depth was found to be $y_{se}/b \cong 3$ for a lower layer of uniform sediment. The increase in y_{se}/b was mainly due to the reduced downstream "support" of the scour hole, i.e. material could escape from the scour hole more readily because of the lower downstream rim.

The third case appears to present the most dangerous case for practice and could occur, for example, when a gravel layer covers a bed of fine sand. A failure of this kind occurred in New Zealand (Bulls bridge in the Rangitikei river, 1973) at a flow rate which could be expected about once a year. A combination of effects, moved location of the main stream channel (general scour) and increased angularity of flow, led to the local scour penetrating the gravel layer ($d_{65} = 20$ mm, $d_{50} = 11$ mm, $d_{35} = 5$ mm) and exposing a bed of fine sand ($d_{50} = 0.15$ mm). The strong eddying around the pier suspended the sand which was then transported from the site in suspension. No compensating inflow of sand from upstream occurred. A gravel ledge formed upstream

Figure 5.16. Photo at collapsed Bulls Bridge site looking upstream. Notice the edge of the intact gravel layer.

and the combined scour depth exceeded 12 m, only a little more than twice the effective width of the pier. Figure 5.16 clearly shows the edge of the intact gravel layer.

An estimate for scour depth H (Figure 5.15(3)) can be made by assuming the upstream bed to be flat, its roughness to be given by the size of gravel and the bed to be at threshold conditions.

A combination of the relationships:

$$q = \frac{1}{n} y_o^{5/3} S^{1/2} \quad n = 0.0417 d_{(m)}^{1/6} \quad \text{and} \quad u_*^2 = g y_o S$$

leads to

$$\left(\frac{u_{*2}}{u_{*1}}\right)^2 = \left(\frac{d_2}{d_1}\right)^{1/3} \left(\frac{y_1}{y_2}\right)^{7/3} \tag{5.8}$$

where d_1 and d_2 are the diameters of the upstream coarse surface particles and the underlying surface fine particles and y_1 and y_2 are the corresponding uniform flow depths over a flat bed of grain roughness at the same slope.

Assuming that $u_{*2} = u_{*2c}$ and substituting $u_{*c} = [\theta_c g \Delta d]^{1/2}$ yields

$$y_2 = y_1 \left\{ \left(\frac{u_{*1}}{u_{*1c}}\right) \left(\frac{\theta_{c1}}{\theta_{c2}}\right)^{1/2} \left(\frac{\Delta_1}{\Delta_2}\right)^{1/2} \left(\frac{d_1}{d_2}\right)^{1/3} \right\}^{6/7} \tag{5.9}$$

When a sill forms upstream of the pier

$$H = \eta(y_2 - y_1) \tag{5.10}$$

The coefficient η for non-ripple-forming sediments from laboratory studies is 1.3 to 2.6. The latter value is recommended for design purposes. For ripple forming sediments, the laboratory value for η is $\eta \cong 1$. However, this value is likely to be an underestimate for field conditions because the turbulence levels are very much higher in the field, particularly when the flow is angled to the pier. Then the fine sands could be easily suspended locally and carried away, because at a pier the flow conditions are substantially different from the two-dimensional flow downstream of a still across the flow.

5.7. LOCAL SCOUR UNDER UNSTEADY FLOW

Unsteady flows occur during such events as the passage of flood waves in rivers, demand surges downstream of hydro-power stations and gravity wave action. These are all waves but of very different wave lengths.

The passage of waves increases the number of variables affecting the local scour. The waves may be translating, like a flood, or oscillatory like the wind waves. The scour may be due to the orbital velocity alone, or the latter may be superimposed on a current. The waves may also cause significant pressure waves in the bed, and these assist in loosening of the bed sediment.

The oscillatory wave on its own leads to a relatively small net scour depth at a cylindrical pile, although the temporary scour with fairly long waves may be significant (Raudkivi 1976). Zanke (1981) introduced the parameter A/b, where $A = H/\sinh kh$ is the double amplitude of the orbital motion of water at the bed and b is the diameter or width of structure. This parameter combines the depth of flow, wave height, wave length and size of structure in a descriptive manner. For values of $A/b > 100$, the conditions of steady flow are approached. With decreasing A/b, the net local scour depth, y_s/b, decreases rapidly and the maximum depth moves to the flanks. The scour depth decreases with increasing b since the half period of the orbital motion becomes too short to transport large volume of sediment. With waves superimposed on a current, the net local scour depth approaches the values due to the current only, i.e. as long as the mean shear stress due to orbital velocities is higher than the critical value.

5.8. SCOUR AT PILE GROUPS

A bridge engineer has direct control over the geometry of his bridge foundations. He should, therefore, have an appreciation of the consequences of using different configurations, such as circular or rectangular piers or pile groups. Scouring at piers has been more thoroughly investigated than that at pile groups. Useful insight to this aspect of local scour is provided by an investigation by Hannah (1978) who studied local scour at groups of cylindrical piles with steady uniform flow and clear-water conditions. A series of tests was first performed on single piles and for a pier (length:width = 6:1) with semicircular ends to provide a basis against which scour at pile groups could be evaluated. Pile groups of various spacings and with different angles of attack were then investigated for one flow condition. For the sediment used in all tests, $d_{50} = 0.75$ mm and $\sigma_g = 1.32$.

Tests showed that scour depths were 80% of equilibrium scour depths after seven hours. Further, after seven hours, only minor changes occurred in scour and deposition patterns. Consequently all tests reported herein were run for seven hours.

5.8.1. Single pile

For a cylindrical pile 33 mm in diameter, the scour depth after seven hours was 62 mm with a velocity of 0.285 m/s (92% of threshold) and a depth of 140 mm. All pile group measurements were made with the same flow velocity and depth so that the results could be compared directly with those of a single pile.

5.8.2. Pile groups

Four of the mechanisms which affect scour at pile groups are not present in scouring at a single pile:

(i) *Reinforcing* causes increased scour depths at the front pile and is a consequence of the dynamic equilibrium which exists in the base of a scour hole when stable conditions have been reached. Bed material is continuously lifted from the base of the hole by the flow which is not, however, capable of removing this material from the scour hole. Should a downstream pile be so placed that the scour holes overlap, then the bed level is lowered at the rear of the upstream scour hole. It is thus easier for the flow to remove material from this hole and it deepens. As pile separation increases, the reinforcing effect decreases and disappears when the maximum bed level between the piles returns to the undisturbed bed level.

(ii) *Sheltering*: The presence of an upstream pile can cause a reduction of the effective approach velocity for downstream piles. This reduction decreases the effect of the "horse-shoe vortex" and thereby reduces scour at downstream piles. A second form of sheltering occurs if material scoured from the upstream pile is deposited on the bed in front of the downstream pile. Flow is then deflected upward from the bed and around the downstream pile. The strength and thus the effectiveness of the "horseshoe vortex" at this pile is thereby reduced. As pile separation increases, the velocity deficit in the wake of the upstream pile disappears and the sheltering effects becomes negligible.

(iii) *Shed Vortices*: Vortices shed from an upstream pile are convected downstream following paths as described in Section 5.2. Should a second pile be placed close to one of these paths, the vortices, by virtue of their velocity and pressure distributions, assist in lifting material from the scour hole. The scouring potential of the shed vortex is a function of the intensity of its convection speed and of the distance between the path and the effected pile. This effect, therefore, decreases more rapidly for piles in line with the flow than for those at angles of attack which place downstream piles on the paths traced by vortices shed by upstream piles.

(iv) *Compressed "Horseshoe Vortex"*: When piles are placed transverse to the flow, each will have, except at very close spacings, its own "horseshoe vortex". As pile spacing is decreased, the inner arms of the "horseshoe vortices" will be compressed. This causes velocities within the arms to increase with a consequent increase in scour depths. This effect will also occur for other (non-zero) angles of attack with its importance for any particular angle being strongly dependent on the pier spacing.

5.8.3. Two piles

In Line (Angle of Attack $0°$). Figure 5.17 shows how the scour depths at each of the piles and the bed level between the piles depend upon the relative spacing a/b, where a is the distance between the centre lines of the piles and b is the common pile diameter.

For two piles touching ($a/b=1$), the scour depth at the front pile is the same as for a single pile d_s, but with increasing separation, the front pile experiences the reinforcing effect which reaches a maximum at $a/b=2.5$ and is evident until $a/b=11$. For larger spacings, the scour depth is the same as for a single pile. With three piles in line at equal spacings up to $a/b=6$, the scour at the middle pile was deeper and at the third pile the scour was shallower than at the downstream pile of the two-pile group.

Transverse (Angle of Attack $90°$). Figure 5.18 shows results of tests in which two piles were set at right angles to the flow. Both piles scoured to the same depth (± 2 mm) and the results have been averaged to define a single curve. The mid-distance scour is also shown.

For the twin pile, $a/b=1$, the scour depth is $1.93 d_s$. This result is in accordance with the concept of scour depth being proportional to the frontal width of the pile. The scour depth decreases rapidly with separation to $1.3 d_s$ at $a/b=1.5$ and with a/b greater than 8, the scour depth is essentially that of a single pile. The scour holes become separate entities at $a/b=11$, and the bed midway between them regains its original level.

At close spacing, ($a/b<2$), the increased scouring results from the increase in effective pile diameter. At $a/b>2$, separate "horseshoe vortices" form. Between the piles, these are compressed thus creating higher velocities and greater scouring potential; this effect reduces as a/b increases and reaches zero as $a/b=8$.

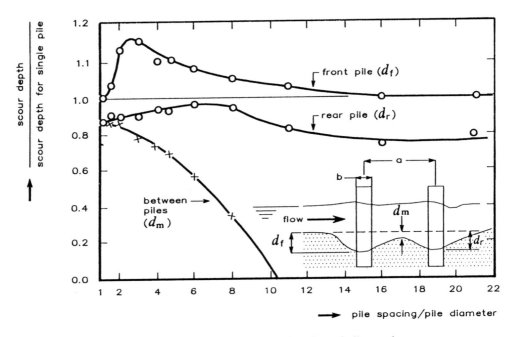

Figure 5.17. Scour depths for two piles in line as a function of pile spacing.

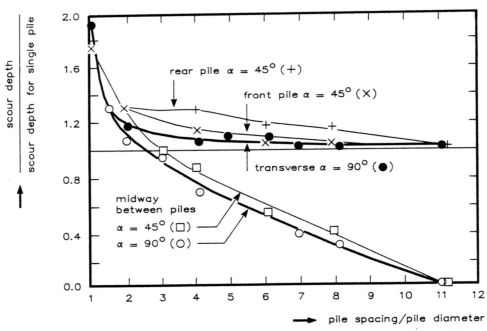

Figure 5.18. Scour depths for two piles as a function of pile spacing. Transverse piles – solid lines. Angle of attack 45° – dashed lines.

Non-aligned (Angle of Attack 45°). Figure 5.18 shows results of tests with two piles having their line of centres at 45° to the approach flow. The twin pile scour depth was $1.77d_s$. This is slightly more than would be suggested by a linear dependence of scour depth on frontal width, namely 1.71 times.

Scour depth at the rear pile exceeded that at a front pile for all values at a/b between 1 and 11. At greater separations, the piles act independently having scour depths equal to that of a single pile and the bed level midway between the piles is unaffected. Increased depths at the rear pile are caused by a combination of the action of the shed vortices from the front pile and compression of the "horseshoe vortices" between the two piles. Evidently, these two processes overcome any sheltering effects and have their maximum influence at $a/b=4$ where the difference between front and rear scour depths was a maximum.

Effect of Angle of Attack, α. Tests with $a/b=5$ were done for $0° < \alpha < 90°$ at 15° intervals with the results shown in Figure 5.19. Scour at the front pile was not very sensitive to angle of attack varying by less than 5% of its value at $\alpha=0°$. Scour depths at the rear pile are much more sensitive to changes in angle of attack as the various scouring mechanisms, described above, come into effect.

At small angles, ($\alpha < 15°$), the dominant effect at the rear pile is sheltering by the front pile. As the angle increases, sheltering is reduced and the pile is affected by shed vortices. Consequently, scour depths increase reaching a maximum (for $a/b=5$) at approximately 45°. Scour depths reduce when the pile moves clear of the shed vortices and approach that of a single pile as the angle of attack approaches 90°.

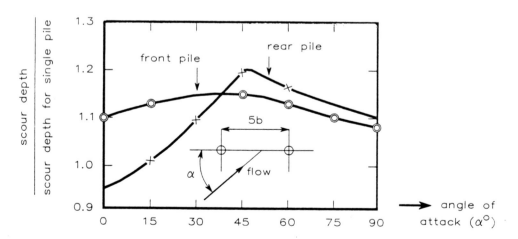

Figure 5.19. The effect of angle of attack, α, on scour depths at two piles spaced five pile diameters apart.

5.9. SUMMARY OF PIER SCOUR ESTIMATIONS

As a general guide the expression

$$y_s = 2.3 K_\alpha b \qquad (5.11)$$

is recommended for a first order estimate of local scour depth in relatively deep flow. Only if flow velocities and direction of the design flood can be predicted with confidence are refinements warranted; even the shape factor is overshadowed by a small angularity of the approach flow, i.e., by K_α.

If the flow can create the transition flat bed conditions, the clear water scour depth $y_s \cong 2.3b$, as for uniform sediment, is again the best estimate, i.e., together with the appropriate coefficients for flow depth, alignment etc. as shown on the flow chart below. However, when flow conditions are not likely to produce the transition flat bed stage the live-bed method of estimation could be used. This approach indicates for a sediment with a broad grain size distribution an appreciably reduced local scour depth.

If the bed is distinctly layered assume that case 3 in Figure 5.15 occurs (the worst case). Calculate H from Equation 5.10. The estimate for the total scour depth is then given by the total scour estimated above plus H, i.e.

$$\text{Total scour depth} = y_{se} + A + B + H$$

Note that the above could be a gross underestimate if the lower layer of sediment is so fine that it will be easily suspended by the increased eddying around the pier.

Scour at piers consisting of two piles is summarized in Section 5.8.3.

The process of estimation of the depth of local scour at a bridge pier can be summarized into a flow chart (see next page).

5.10. NUMERICAL EXAMPLES

1. Estimate the maximum depth of local clear-water scour at a pier of diameter 1.2 m in a depth of flow of 1.5 m when the bed material is

(a) $d_{50} = 0.5$ mm, $\sigma_g = 2.5$
(b) $d_{50} = 100$ mm, $\sigma_g = 2.5$

From Equation 5.3

$$y_{se}(\sigma) = 2.3 K_\sigma b$$

where from Figure 5.5, the value of $K_\sigma = 0.42$ in both cases, i.e.

$$y_{se}(\sigma) = 2.3 \times 0.42 \times 1.2 = 1.16 \text{ m}.$$

For $b/d_{50} = 1.2/0.0005 = 2400$ and $1.2/0.1 = 12$, and $y_o/b = 1.25$ the relative grain size effect from Figure 5.6 is $K(b/d_{50}) = 1.0$ and 0.85, respectively. From Figure 5.7 the flow

Scour at bridge piers 89

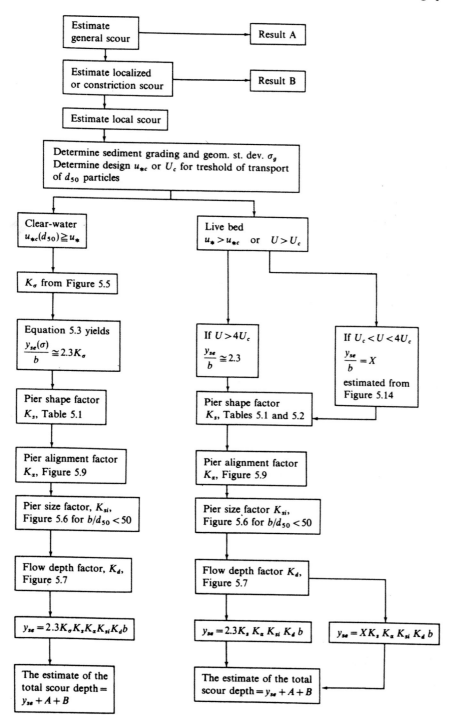

Flow chart for pier scour estimation.

depth effect is $K_d = 0.84$ and 0.55, respectively. Hence, $y_{se} = 1.16 \times 0.84 = 0.97$ m and $y_{se} = 1.16 \times 0.85 \times 0.55 = 0.54$ m, respectively.

2. Estimate live-bed scour at a 2.0 m diameter pier in a flow of 2.0 m depth just upstream of the scour hole. Slope is 1:200 and the sediment parameters are: $d_{50} = 20$ mm, $d_{max} = 150$ mm, $\sigma_g = 3.5$ and specific gravity 2.6.

The clear-water peak depth of scour for *uniform particle size* is

$$y_{se} = 2.3 \times 2.0 \times 0.81 = 3.7 \text{ m}$$

where the factor 0.81 is from Figure 5.7 for $y_o/b = 1$. The expected scour depth at transition flat bed conditions would be the same if the scour hole does not armour.

From Equation 5.6 the limiting armour size is $d_{50a} = 83$ mm for which, from the Shields' relationship $u_{*c}^2/\Delta gd \cong 0.056$, the limiting armour shear velocity $u_{*a} = 0.27$ m/s. The corresponding mean velocity from Equation 5.7 is $U_{ca} = 3.30$ m/s. The shear velocity based on d_{50} is $u_{*c} = 0.13$ m/s and the corresponding mean critical velocity for initiation of sediment transport $U_c \cong 2.05$ m/s. The shear velocity based on d_{50} and the slope $u_* = (g \times 2 \times 0.005)^{1/2} = 0.31$ m/s and $U = 4.89$ m/s, i.e., $U/U_c = 2.4$. For this ratio Figure 5.14 indicates $y_{se}/b \cong 1.27$ or $y_{se} = 2.54$ m in deep water. Figure 5.7 with $y_o/b = 1$ and $b/d_{50} = 100$ yields $K_d = 0.8$. Thus,

$$y_{se} = 2.0 \text{ m}$$

However, since the local scour could also occur at lower flow rates, corresponding at armour peak conditions, this value cannot be used for design. The design local scour depth would be that at the armour peak, $y_{se}/b \cong 2$, which after correction for relative water depth leads to a design value for local scour of $y_{se} \cong 3.2$ m.

No preferential armouring of the scour hole is likely since the large stones, d_{max}, are smaller than $0.1b = 200$ mm.

3. Estimate the live-bed scour depth for the above example when pier diameter is 1.0 m. The maximum clear-water scour depth for uniform sediment is

$$y_{se} = 2.3 \times 1 \times 0.9 = 2.1 \text{ m}$$

Now $d_{max} > 0.1b = 100$ mm. For the large stones $b/d = 1/0.15 = 6.7$ and according to Figure 5.6 the scour depth is reduced by a factor of about 0.6, i.e., $y_{se} \cong 1.2$ m. This reduction arises from the accumulation of the large stones in the scour hole. These armour the hole as well as allow the down flow to penetrate into the bed and thus dissipate some of its energy.

4. A gravel layer of $d_{50} = 10$ mm covers a bed of

(1) sand: $d_{50} = 0.30$ mm, $\sigma_g = 1.5$ and $\rho_s = 2650$ kg/m^3
(2) gravel: $d_{50} = 3.0$ mm, $\sigma_g = 1.5$ and $\rho_s = 2650$ kg/m^3

During a flood, the local scour at a 2 m diameter pier penetrates the gravel layer which remains intact upstream, forming a sill. Estimate the depth of scour below the original bed level when the slope of the river $S=0.001$.

From Shields' diagram, $\theta_c = 0.056$ for $d_{50} = 10$ and 3 mm and $\theta_c = 0.034$ for $d_{50} = 0.30$ mm. Hence, assuming that the 10 mm gravel layer is at threshold on the bed upstream of the pier:

$$(u_{*1})^2 = 0.056 \times 1.65 \times 9.81 \times 0.01 = 0.0091$$
$$u_{*1} = 0.095 \text{ m/s} = (gy_1 S)^{1/2}$$
$$y_1 = 0.924 \text{ m}.$$

For $d_{50} = 0.30$ mm:

$$y_2 - y_1 = 0.924 \left\{ \left[\left(\frac{0.056}{0.034}\right)^{1/2} \left(\frac{1.0}{0.3}\right)^{1/3} \right]^{6/7} - 1 \right\} = 2.19 \text{ m}$$

The depth of local scour in 0.30 mm sand (Equation 5.3) in deep water is

$$y_{se}(\sigma) = 1.0 \times 2.3 \times 2.0 = 4.60 \text{ m}.$$

This scour is in about 3.11 m water, i.e., $y_o/b = 3.11/2 = 1.55$ and $b/d_{50} = 6667$. These yield from Figure 5.7 an adjustment factor of $K_d = 0.88$ and $y_{se} = 4.22$ m. The total scour depth is then of the order of $4.2 + 2.2 = 6.4$ m. This value with fine sands may be affected by additional scouring due to localised high turbulence at the pier.

For $d_{50} = 3.0$ mm:

$$y_2 - y_1 = 0.924 \left\{ \left[\left(\tfrac{10}{3}\right)^{1/3} \right]^{6/7} - 1 \right\} = 0.38 \text{ m}.$$

$$H = 2.6 \times 0.38 = 1.0 \text{ m}$$

The local scour depth of 4.6 m is modified by grading, $K_\sigma \cong 0.92$ and by depth, $y_o/b = 1.92/2 = 0.96$ and $b/b_{50} = 667$, $K_d \cong 0.8$ to

$$y_{se} = 0.92 \times 0.8 \times 2.3 \times 2.0 = 3.4 \text{ m}$$

The total scour depth below the undisturbed upstream bed level is of the order of 4.4 m.

5.11. SCOUR PROTECTION

The idea of bed protection and prevention of scour at a pier has attracted a good deal of attention. Reduction of scour depth would mean shallower foundations and reduced cost.

The most common method of scour "prevention", a riprap protection, is the dumping of stones on the river bed around the pier, or dumping stones into the scour hole around the pier. A rule of thumb is that the width of the protection measured from the pier should be about three to four times the width (projected) of the pier. Bonasoundas (1973) recommended a riprap protection for a cylindrical pier in the shape of longitudinal cross section of an egg, with the blunt end facing the flow. The overall recommended width is $6b$ and length $7b$ of which $2.5b$ is upstream of the upstream face of the pier. The recommended thickness of riprap is $b/3$ and minimum stone size is

$$d(\text{cm}) = 6 - 3.3U + 4U^2 \qquad (5.12)$$

where U is in m/s.

In general, scour starts at the pier at about half the threshold velocity of the sediment on the upstream bed. Hence, the critical or design velocity for the riprap is

$$2U < U_c \qquad (5.13)$$

where U is the mean approach flow velocity at design discharge.

The combination of Strickler and Manning formulae yields

$$\frac{U}{u_*} = 7.66 \left(\frac{R}{d}\right)^{1/6}$$

and the beginning of the movement is given by Shields criterion or by Figure 2.2. For coarse bed material a conservative value is

$$\frac{u_{*c}^2}{\Delta g d} \cong 0.04$$

Thus, for a two-dimensional flow with R equal to flow depth y_o the relationship between the mean critical flow velocity and armour stone size is

$$U_c = 4.8 \Delta^{1/2} d^{1/3} y_o^{1/6} \cong 6 d^{1/3} y_o^{1/6} \qquad (5.14)$$

for stones with density $\rho_s = 2600$ kg/m^3 and d measured in m.

The stone sizes given above are generally smaller than by the various codes of practice which incorporate safety factors.

The threshold approach to rip-rap stability is affected not only by the choice of the critical threshold value of θ, but also by the shape of the stones. The drag coefficient for the drag force exerted by the fluid varies with the shape and the roughness of the stones.

The simplest empirical relationships between mean velocity and rock size is

$$U_c = 4.92 \sqrt{d} \qquad (5.15)$$

where U_c is in m/s, d is the diameter of the equivalent sphere in m, and the density $\rho_s = 2600$ kg/m^3. This applies for a horizontal bed. Most empirical relationships give d proportional to U_c^2 rather than U_c^3 which follows the observation by Brahms in 1754 that U_c is proportional to (mass)$^{1/6}$, i.e. $U_c^2 \propto d$. The reduction in stability on a slope has to be added.

The stability of the riprap protection is strongly affected by the roughness of the surface on which the rocks rest as illustrated by the empirical formula for the angle of repose

$$\tan \phi = 1/(0.3 + 0.59 d/k) \qquad (5.16)$$

where d is the riprap "size" and k is the size (roughness) of the layer on which the riprap unit rests. This dependence is one of the reasons for the requirement of at least two layers of armour rock. The stability of the larger units in the surface layer is also affected by the size of the neighbouring rocks. These effects are accounted for by empirical safety factors.

The relationship given by the Shore-Protection Manual (1973) is

$$\frac{U_c}{\sqrt{2g}} = Y\Delta^{1/2}(\cos\alpha - \sin\alpha)^{1/2}d^{1/2} \tag{5.17}$$

where α is the slope in flow direction, Y is a constant equal to 1.20 for embedded rock and 0.86 for nonembedded rock (Isbash constant), and

$$d_{(m)} = 1.24\,(\text{volume})^{1/3}$$

For a horizontal bed, $\rho_s = 2600$ kg/m³ and $Y = 0.86$, Equation 5.17 leads to $U_c = 4.82\sqrt{d}$.

The riprap should be composed of a well-graded mixture of rocks so that the voids between large stones are filled by smaller ones and the stones support each other. The thickness should be two layers of riprap or more. The rocks are characterized by a representative size

$$d_r = \left[\frac{1}{10}\sum_{i=1}^{10} d_i^3\right]^{1/3} \tag{5.18}$$

where $d_i = 1/2(d_o + d_{10})$; $d_2 = 1/2(d_{10} + d_{20})$ etc.

The d_r-size is approximately equal to d_{67} by weight. An illustration of a well-graded rip-rap is given by the following values:

d/d_{50}	0.2	0.33	0.5	0.7	1.0	1.15	1.5	2
% by weight	0	10	16	30	50	60	84	100

In addition riprap should be placed on a suitable inverted filter or a geotextile filter. The filter layer protects the bed or bank material from being winnowed away by the currents and the riprap protects the filter. Without proper filter the protection may disintegrate through loss of soil from underneath.

Riprap is more likely to be successful if the river bed level does not vary much, that is if the trough elevations of largest bed features can be estimated with confidence, and the riprap can be placed at the trough elevation, as shown in Figure 5.20.

Where the bed is subject to substantial lowering during a flood, the riprap protection could be destroyed by undermining from the sides as the river bed around the protection is lowered. If the riprap is piled high around the pier, it may act as part of a pier of larger diameter and actually aggravate conditions. Neill (1964) reported a case where "the deep-water piers had been surrounded by large heaps of stone extending up to nearly low-water stage, about 10 m above the bed. One pier collapsed in a flood, apparently as a result of undermining from downstream", Such a footing presents a much wider "pier" to the flow and leads to more severe erosion.

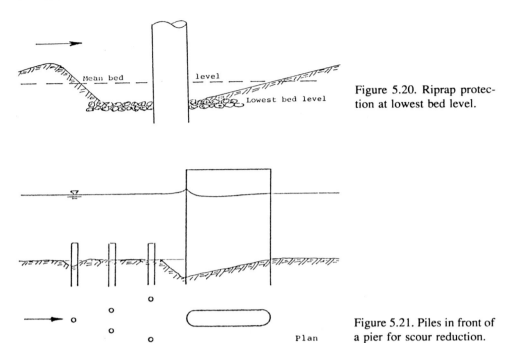

Figure 5.20. Riprap protection at lowest bed level.

Figure 5.21. Piles in front of a pier for scour reduction.

Riprap protections become useless in rivers with substantial internal stream channels which move about as, for example, in braiding rivers.

Chabert and Engeldinger (1956) investigated the effectiveness of piles in front of a pier for scour reduction, Figure 5.21. The effectiveness is a function of the number of piles, their protrusion and spacing from each other and from the pier, and the angle made by the two lines of piles. The authors reported up to 50% reduction in scour depth but gave no design criteria other than the results of model tests. Again, such a protection is of value only if the general bed level during a flood can be predicted with confidence, and the bed level movements are relatively small. The design has to be based on model tests.

5.12. FIELD MEASUREMENT OF SCOUR DEPTH

Field data on scour at bridge piers during floods are extremely difficult to collect and as a consequence such data are rare. The need for the data was pointed out by the U.S. Highway Research Board in stating that "the first priority in research on scour problems should be given to field measurements".

Various methods have been used to collect field data. Sanden (1960) appears to have been the first to use an echo-sounder to measure scour depth. The method uses a narrow beam transducer which may be mobile and operated from a bridge deck or

Scour at bridge piers 95

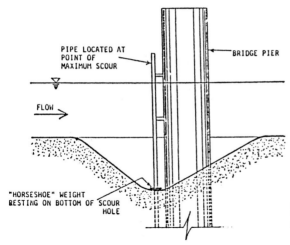

Figure 5.22. Schematic illustration of scour measuring device.

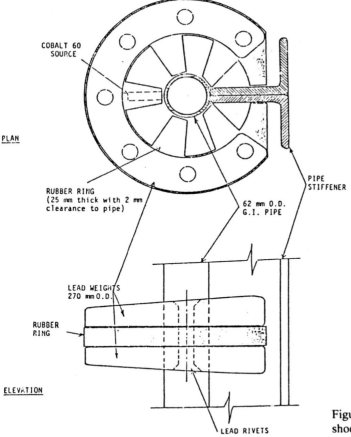

Figure 5.23. Details of "horseshoe" weight.

a boat, or permanently fixed to the bridge structure. One of the main difficulties inherent in the method is that high concentrations of sediment in suspension scatter and absorb the sonic pulses, preventing reception of a detectable or quantifiable echo. The instrument is also complex and requires a power supply for permanent installation. The site may not be accessible by mobile equipment during a flood.

Other methods include soundings by weighted wire, electrical resistivity method (Hubbard 1955) and geotechnical methods. The latter aim at deduction of previous maximum scour depths after a flood has subsided (Kühn and Williams 1961). These depend on detecting a discontinuity in the soil profile of the scour bed, indicating the maximum depth. The most important measurement for a bridge designer is the maximum scour depth during a given flood. With this in mind a method of measurement of maximum scour depth was developed at the University of Auckland, New Zealand. The instrument uses a portable gamma spectrometer to detect the elevation of a small radioactive source (Cobalt 60). The source is housed in a horseshoe-shaped lead weight which slides down a pipe embedded in the scour hole, Figure 5.22. The pipe is designed and supported to withstand local flow conditions and its bottom end is sealed. The horseshoe resting on the bed of the initial scour hole will slide down the pipe as the scour hole deepens during a flood and comes to rest at the bottom of the scour hole produced. As the flood subsides the horseshoe becomes buried. The elevation of the maximum scour depth for the known flood can then be measured subsequently by lowering the gamma-ray probe down the inside of the pipe. The level of the horseshoe is indicated by a maximum on a meter or an audible signal. The horseshoe is illustrated in Figure 5.23 and the measuring system in Figure 5.24.

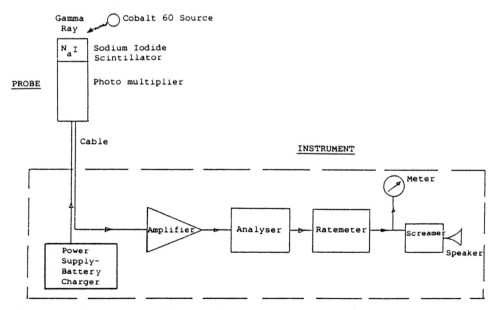

Figure 5.24. Block diagram of the portable Gamma Spectrometer.

REFERENCES

Anderson, A.G. 1974. *Scour at bridge waterways – A review*. U.S. Dept. of Transportation. Office of Research and Development, Environmental Design and Control Division.

Baker, R.E. 1986. *Local Scour at Bridge Piers in Non-Uniform Sediments*. University of Auckland, N.Z., Dept. Civil Engng, Report No. 402.

Bonasoundas, M. 1973. *Strömungsvorgang und Kolkproblem*. Oscar von Miller Institut, Techn. Univ. Munich, Rep. No. 28.

Breusers, H.N.C. 1965. Scour around drilling platforms. *Hydraulic Research 1964-65*, IAHR 19; 276.

Breusers, H.N.C., G. Nicollet & H.W. Shen 1977. Local scour around cylindrical piers. *J. Hydr. Res.* 15; 211–252.

Chabert, J. & P. Engeldinger 1956. *Etude des afouillements autour des piles des ponts*. Laboratoire National d'Hydraulique, Chatou, France.

Chee, R.K.W. 1982. *Live-bed scour at bridge piers*. Report No. 290 University of Auckland, School of Engineering, New Zealand.

Chiew, Y.M. 1984. *Local Scour at Bridge Piers*. Rep. No. 355, University of Auckland, School of Engineering, New Zealand.

Chin, C.O. 1985 *Stream bed armouring*. Eng. Rep. No. 403, Univ. of Auckland, N.Z., Dept. Civil. Engng.

Dietz, J.W. 1972. Systematische Modellversuche über die Pfeilerkolkbildung. *Mitteilungsblatt der Bundesanstalt für Wasserbau*, Nr. 31; 95–109.

Dietz, J.W. 1972. Modellversuche über Kolkbildung. *Die Bautechnik* 49; 162–168, 240–245.

Dietz, J.W. 1973. Kolkbildung and einem kreiszylindrischen Pfeiler. *Die Bautechnik* 50; 203–208.

Ettema, R. 1980. *Scour at bridge piers*. Univ. of Auckland, N.Z., School of Engineering, Rep. No. 216.

Hannah, C.R. 1978. *Scour at pile groups*. University of Canterbury, N.Z., Civil Engng., Research Rep. No. 78-3, 92 pp.

Hjorth, P. 1972. *Lokal Erosion och Erosionsverken vid Arloppsledning i Kustnära Omräden*. Inst. Vattenbyggnad, Tekn. Högskolan; Lund, Bulletin Serie B, Nr. 21.

Hjorth, P. 1975. *Studies on the nature of local scour*. Dept. Water Res. Engng., Lund Inst. of Technology, Bulletin Series A, No. 46.

Hopkins, G.R., R.W. Vance & B. Kasraie 1975. *Scour around bridge piers*. Report No. FHWA-RD-75-56 or Report No. FHWA-RD-78-103 1980. Final Report prepared for Federal Highway Administration, Washington D.C.

Hubbard, P.G. 1955. Field measurement of bridge pier scour. *Proc. Highway Research Board* 34; 184–188.

Kühn, S.H. & A.A.B. Williams 1961. Scour depth and soil profile determination in river beds. *Proc. 5th Int. Conf. Soil Mech. and Found. Eng., Paris*; 487–490.

Laursen, E.M. & A. Toch 1956. *Scour around bridge piers and abutments*. Iowa Highway Res. Board, Bulletin No. 4, 60 pp.

Melville, B.W. 1974. *Scour at bridge sites*. Univ. of Auckland School of Engineering, New Zealand, Report No. 104 or Proc. National Roads Board, RRU Seminar on Bridge Design and Research.

Melville, B.W. 1975. *Local scour at bridge sites*. Univ. of Auckland, N.Z., School of Engineering Rep. No. 117.

Melville, B.W. & A.J. Raudkivi 1977. Flow characteristics in local scour at bridge piers. *J. Hydr. Res.* 15; 373–380.

Neill, C.R. 1964. *River-bed scour, a review for engineers*. Canadian Good Roads' Assoc. Techn. Publ. No. 23, Ottawa.

Neill, C.R. (ed.) 1973. *Guide to bridge hydraulics.* University of Toronto Press, Toronto.
Raudkivi, A.J. 1976. *Loose boundary hydraulics,* 2nd Edition. Pergamon Press, Oxford.
Raudkivi, A.J. 1981. *Grundlagen des Sedimenttransports.* Sonderforschungsbereich 79, Univ. Hannover.
Raudkivi, A.J. & R. Ettema 1977a. Effect of sediment gradation on clear-water scour and measurement of scour depth. *Proc. 17th Congress IAHR, Baden-Baden* 4; 521–527.
Raudkivi, A.J. & R. Ettema 1977b. Effect of sediment gradation on clear-water scour. *Proc. ASCE* 103(HY10); 1209–1213.
Raudkivi, A.J. & A.J. Sutherland 1981. *Scour at bridge crossings.* National Roads Board, Road Research Unit Bulletin No. 54, 100 pp., Wellington, New Zealand.
Sanden, E.J. 1960. Scour at bridge piers and erosion of river banks. Western Assoc. of Canadian Highway officials. *Proc 13th Ann. Conf.*
Tison, L.J. 1940. Erosion autour de piles de pont en rivière. *Ann. des Travaux Publics de Belgique* 41; 813–871.
Zanke, V. 1981. *Seegang erzeugte Kolke am Bauwerken.* Sonderforschungsbereich 79, Teilprojekt B9, TU Hannover.

CHAPTER 6

Scour by jets, at high-head structures and at culvert outlets

H.N.C. BREUSERS

6.1. INTRODUCTION

Flows through hydraulic structures often issue in the form of jets. The velocities are usually high enough for these jets to produce sizable, even dangerous scour holes. The jets can occur in many different configurations; studies include horizontal and vertical jets, two and three dimensional jets, submerged or free jets and jets with various boundary configurations. Scour downstream of culverts is a common and rather special case. These phenomena are treated in this chapter; scour by flows from low-head structures is discussed in Chapter 7.

Because of the variety of cases studied, a large number of symbols is required. Symbols are frequently defined with reference to figures representing both the phenomenon and its characteristics. For the general list of symbols see pages VII-VIII.

6.2. CHARACTERISTICS OF SUBMERGED JETS

Fully submerged jets, both plane or two-dimensional (2D) and circular (3D) are characterized by a linear increase in width and a Gaussian velocity distribution in the fully developed part of the jet. The first part of the jet, the flow establishment zone, has a core of diminishing width and length L_k, within which the velocity is equal to the exit velocity U_o, see Figure 6.1.

The main characteristics for fully developed jet flow are (Rajaratnam 1976):

— decrease of maximum velocity U_m for $x > L_k$:

$$\text{circular jet:} \quad \frac{U_m}{U_o} = A_3 \frac{D_u}{x} \tag{6.1a}$$

$$L_k = A_3 D_u$$

$$\text{plane jet:} \quad \frac{U_m}{U_o} = \left(A_2 \frac{2B_u}{x}\right)^{1/2} \tag{6.1b}$$

$$L_k = A_2(2B_u)$$

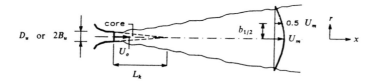

Figure 6.1. Definition sketch for jets.

— increase in halfwidth $b_{1/2}$ (point where $U/U_m = 0.5$):

$$\text{circular jet: } \beta_3 = \frac{b_{1/2}}{x} = \frac{0.589}{A_3} \qquad (6.2\text{a})$$

$$\text{plane jet: } \beta_2 = \frac{b_{1/2}}{x} = \frac{0.664}{A_2} \qquad (6.2\text{b})$$

— velocity distribution for both jet types:

$$\frac{U}{U_m} = e^{-k(r/x)^2}$$

$$\text{circular jet: } k_3 = 2A_3^2 \qquad (6.3\text{a})$$

$$\text{plane jet: } k_2 = \frac{\pi}{2} A_2^2 \qquad (6.3\text{b})$$

Values of constants quoted vary between authors, but a commonly accepted value is $A_2 = A_3 = 6.0$ which leads to:

$\beta_3 = 0.098 \quad k_3 = 72$

$\beta_2 = 0.11 \quad k_2 = 56$

The discharge of the jet increases owing to the entrainment of surrounding water:

$$\text{round jet: } \frac{Q}{Q_o} = 0.33 \frac{x}{D_u} \qquad (6.4\text{a})$$

$$\text{plane jet: } \frac{Q}{Q_o} = 0.57 \left(\frac{x}{2B_u}\right)^{1/2} \qquad (6.4\text{b})$$

A jet which issues parallel to a wall, with its origin at the wall, is called a *wall jet*. Its main characteristics in the outer zone are similar to those of a free jet. Along the wall a relatively thin, almost negligible boundary layer develops. Relations for U_m and $b_{1/2}$ from free jets are generally valid for wall jets with minor changes in the constants according to Rajaratnam (1976).

If a fully submerged circular jet strikes a wall perpendicularly, a stagnation zone with increased pressure is formed. The maximum pressure at the wall occurs at the jet axis and is given by:

$$P_{max} = 50 \, (1/2\rho U_o^2)\left(\frac{D_u}{H}\right)^2 \tag{6.5}$$

A comparison with Equation 6.1a indicates that the effective value of $x = 0.85H$ for $A_3 = 6$. Accordingly U_m depends not upon H but only upon the kinematic momentum flux

$$M = Q_o U_o \tag{6.6}$$

in which Q_o = jet discharge.

Beyond the impact zone a radial wall jet is formed, see Figure 6.2, with maximum velocity (Rajaratnam 1976)

$$\frac{U_m}{U_o} = 1.03 \frac{D_u}{r} \tag{6.7}$$

Equation (6.6) can be transformed into

$$U_m = 1.16 \frac{M^{1/2}}{r} \tag{6.8}$$

For submerged unbounded jets U_m can also be expressed in terms of M.

round jet: $\quad U_m = A_3 \left(\frac{\pi}{4}\right)^{1/2} \frac{M^{1/2}}{x}$

or $\quad U_m = 6.8 \dfrac{M^{1/2}}{x} \tag{6.9}$

plane jet: $(M = q_o U_o$ in which q_o is jet discharge per m)

$$U_m = A_2^{1/2} \frac{M^{1/2}}{x^{1/2}}$$

or $\quad U_m = 2.45 \dfrac{M^{1/2}}{x^{1/2}} \tag{6.10}$

The foregoing résumé of the behaviour of submerged jets is useful in the interpretation

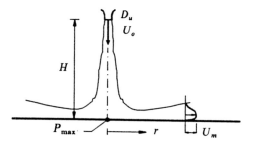

Figure 6.2. Definition sketch radial wall jet.

6.3. SCOUR BY SUBMERGED HORIZONTAL JETS

Laursen (1952) measured the scour by a fully submerged plane jet with $2B_u = 7.6$ mm. The scoured material was sand ($d_{50} = 0.27$, 0.7 and 1.6 mm). Scour depths increased with log t, and the scour rate increased with U_o/w in which w is the fall velocity of the sand. For definitions see Figure 6.3.

Vanoni (1975) analysed test results by Laursen and similar test results by Tarapore (1956). The resulting graphical presentation of scour length seriously underestimates the final extent of scour as was found by Whittaker (1984).

An upper limit for L_{st} is given by Whittaker

$$\frac{L_{st}}{2B_u} = 35 \left(\frac{U_o}{w}\right)^{0.57} U_o^{0.86} \tag{6.11}$$

(constant is dimensional, metric units).

Clarke (1962) performed tests with round horizontal jets ($D_u = 2.4$, 4.8 and 14.3 mm) and sand ($d_{50} = 0.82$ and 2.02 mm).

Scour holes were found to be similar in shape

$$\frac{B_{sm}}{L_s} = 0.57 \pm 0.03 \quad \frac{y_{sm}}{L_s} = 0.15 \pm 0.01 \tag{6.12}$$

where the index m indicates the equilibrium value of y_s.

Scour length L_s was described by

$$\frac{L_s}{D_u} = 4.3 \left(\frac{U_o}{\sqrt{gD_u}}\right)^{7/15} \left(\frac{U_o}{w}\right)^{1/5} \left(\frac{gt}{w}\right)^{1/15} \tag{6.13}$$

Altinbilik and Basmaci (1973) repeated Laursen's tests, deriving for the equilibrium scour depth

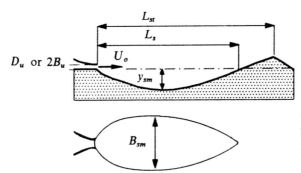

Figure 6.3. Definition sketch for scour by submerged horizontal jet. The lower part of the sketch is for jets of finite width.

$$\frac{y_{sm}}{2B_u} = (\tan \phi)^{0.5} \left(\frac{d}{2B_u}\right)^{0.25} \left(\frac{U_o}{\sqrt{\Delta gd}}\right)^{1.5} \tag{6.14}$$

where

ϕ = angle of repose

$\Delta = (\rho_s - \rho)/\rho$

The experiments were run with $2B_u = 6$ to 50 mm; the bed materials were sand $d = 1.2$ and 6.5 mm and tuff $d = 2.65$ mm ($\rho_s = 1300$ kg/m^3). The velocities were $U_o = 0.6$ to 4.3 m/s. Scour developed rapidly. More than 90% of the final scour depth was attained in about 10 minutes.

Rajaratnam and Berry (1977) repeated the tests by Clarke using round air and water jets of diameter $D_u = 25$ mm in sand and polystyrene, with jet velocities of $U_o = 1.2$–1.8 m/s for sand, $d = 1.4$ mm, in water and 10–54 m/s in air for the polystyrene, $d = 1.4$ mm and $\rho_s = 1041$ kg/m^3. The equilibrium depth can be described, within the range of the experimental data for ($2 < U_o/\sqrt{\Delta gd} < 14$), by

$$\frac{y_{sm}}{D_u} = 0.4 \left\{\frac{U_o}{\sqrt{\Delta gd}} - 2\right\} \tag{6.15}$$

The average value of the length-depth ratio was 5, the average width-depth ratio was 2. The wall jet was found to behave as an unbounded jet up to the point of maximum scour depth. Velocity distributions were Gaussian with $b_{1/2} = 0.097x$.

Rajaratnam (1981a) performed similar tests with plane horizontal jets over sand with $d = 1.2$ and 2.4 mm, and for $2B_u = 1.4$, 2.6 and 9.8 mm. The average value of the ratio L_{st}/y_{sm} was found to be equal to 5. If the jet was not fully submerged the scour dimensions were changed; for a low tail water level equal to $2B_u$, the depths were approximately 50% smaller and the lengths were some 50 to 25% larger (Rajaratnam and MacDougall 1983).

6.4. SCOUR BY SUBMERGED VERTICAL JETS

Doddiah (1953) reported tests with a round jet, for which $D_u = 50$ mm, impinging on a gravel bed ($d = 6$–12 mm and 3–6 mm). Scour generally increased with increasing tailwater depth (0.05–0.4 m) for a fixed position of the jet above the bed level (0.7 m). For small tailwater depths, the jet created a narrow deep hole filled with recirculating grains, giving only a relatively small net scour depth as the grains settled within the scour hole once flow ceased. Scour depths followed a relationship similar to Equation 6.15. A hollow jet, having the same net surface area and jet velocity, produced the same scour depth as the solid round jet.

Clarke (1962) studied the action of a round vertical submerged jet ($D_u = 2.4$, 3.0, 3.8 and 14.3 mm) in sand ($d_{50} = 0.44$, 0.82, 1.27 and 2.0 mm). He noted that the dynamic

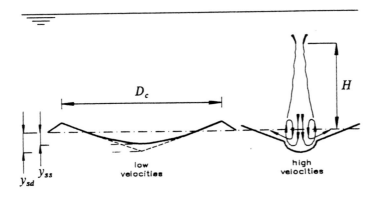

Figure 6.4. Definition sketch for submerged vertical jets.

scour depth, y_{sd}, was larger than the (static) scour, left after the flow ceased, y_{ss}, see Figure 6.4. At higher jet velocities more particles went in suspension giving a hemispherically shaped core.

The value of H had some effect on y_{sd} for small jet pressures but not for large pressures. Scour holes were generally similar with $y_{sd} = (0.21 \pm 0.003) D_c$, where D_c is defined in Figure 6.4. The results were linked by

$$\frac{D_c}{D_u} = 5.5 \left(\frac{U_o}{\sqrt{gD_u}}\right)^{0.43} \left(\frac{U_o}{w}\right)^{0.05} \left(\frac{gt}{w}\right)^{0.05} \tag{6.16}$$

Scour developed very rapidly and 70% of the final depth was reached in a few seconds.

Johnson (1967) compared the scour in gravel ($d_{50} = 10$ mm) due to a solid vertical round jet (A) ($D_u = 40$ mm) with a jet having the same area but divided into 32 jets (B) and a jet with 50% air, having the same water flow rate and velocity (C). Differences between A and B were small but the scour depth for C was much less than that for the others. Scour depths after stopping the jet were also much less than observed during the tests.

Westrich and Kobus (1973), Westrich (1974) and Kobus et al. (1979) measured the scour caused by vertical round jets with constant and variable jet velocity ($D_u = 20$, 30 and 40 mm) in sand with $d_{50} = 1.5$ mm. The distance between jet exit and sand bed varied between 0 and 0.82 m. The form of the scour depended on $K = (U_m/w)^2$ in which U_m is the maximum velocity computed at the sand bed surface. Figure 6.5 shows typical forms of the scour.

The value of H affected somewhat the volume scoured. For given jet parameters, the scour volume first increases with H, and then remains constant before decreasing again, see Figure 6.6.

A sinusoidal variation of the jet velocity generally resulted in a greater volume of scour. The maximum ratio of scoured volume in this situation, was 4 to 5 times greater

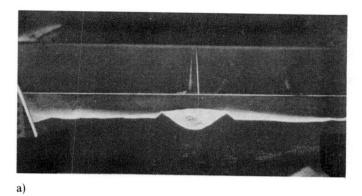

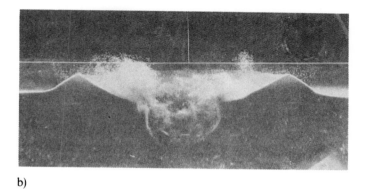

Figure 6.5. Scour forms:
a) $1.5 < K < 3$,
b) $K > 6.5$.

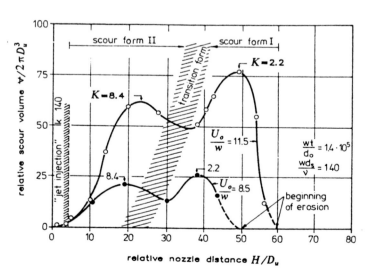

Figure 6.6. Effect of H on scoured volume.

than that for steady flow conditions with same average flow rate.

Rajaratnam (1981a) described tests with vertical plane jets of water ($2B_u = 2.5$ mm) in sand of $d = 1.2$ and 2.38 mm as bed material. His results can be described by

$$\frac{y_{ss}}{2B_u} = 0.23 \left(\frac{U_o}{\sqrt{\Delta g d}}\right) \left(\frac{H}{2B_u}\right)^{1/2} \tag{6.17}$$

The average ratio of D_c/y_{ss} was found to be equal to 5.5.

The ratio of dynamic to static scour depth, y_{sd}/y_{ss} reached values between 1.2 and 2.

Akashi and Saito (1984) performed similar tests with submerged plane jets ($2B_u = 8$, 10, 16 and 20 mm) in sand ($d = 0.28$, 0.77 and 1.41 mm) and with variable stand-off distance. For a discussion of their results see below.

The influence of jet angle was studied for plane submerged jets by Rajaratnam (1981b). Changing the angle between jet axis and sand bed surface did not alter y_{ss} or the total width of the scour hole, the latter did become more asymmetrical, however.

Rajaratnam (1982a) reported tests with a round jet ($D_u = 9.8$ mm) impinging on a sand bed with $d = 1.2$ and 2.38 mm. He also presented some data obtained by Kobus et al. (their scour type (a) results). The data for static scour can be described by

$$\frac{y_{ss}}{D_u} \cong 0.3 \frac{U_o}{\sqrt{\Delta g d}} \tag{6.18}$$

The value of y_{sd}/y_{ss} was found to increase from 1.0 to 1.5 as U_o increased. The average value of D_c/y_{ss} was 4.2.

Rajaratnam (1981c) carried out tests with a low tail water level (partly submerged round jet) and jet sizes of $D_u = 9.8$ and 12.7 mm in sand beds with $d = 1.0$, 1.15 and 2.38 mm. The representative jet parameters U'_o and D'_u were computed at the water surface, i.e.

$$U'_o = (U_o^2 + 2g\Delta H)^{1/2} \tag{6.19}$$

where ΔH is the distance between jet exit and water surface, and a linear relation of the type of Equation 6.18 was obtained

$$\frac{y_{ss}}{D_u} = 0.13 \frac{U'_o}{\sqrt{\Delta g d}} \tag{6.20}$$

The ratio of dynamic to static scour depth, y_{sd}/y_{ss}, was large and reached values in the range 2 to 4.

Rajaratnam (1982b) carried out tests for partly submerged plane jets (low tail water level), using a jet with $2B_u = 2.54$ mm in sand beds with $d = 1.2$ and 2.38 mm. The representative jet parameters U'_o and B'_u were taken at the level of the original uneroded bed. A linear relation was obtained

$$\frac{y_{ss}}{2B_u} = 1.82 \frac{U'_o}{\sqrt{\Delta g d}} \quad \text{for} \quad 17 < \frac{U'_o}{\sqrt{\Delta g d}} < 26$$

The ratio of dynamic to static scour depth, y_{sd}/y_{ss}, varied from 1.1 to 1.3. For equal jet and sand bed characteristics, dynamic scour depth, y_{sd}, was equal for submerged and partly submerged jets for the same distance between jet origin and sand bed. The values of the static scour depths, y_{sd}, were up to 1.7 times larger for the situation with the low tail water level.

6.5. DISCUSSION OF RESULTS ON SCOUR BY SUBMERGED JETS

The simplest approximation for the dimensions of scour by *submerged horizontal jets* is one based on the assumption that the length of the scour hole L is determined by the condition that U_m no longer exceeds U_c, the velocity required for sediment movement. The experiments show that the ratio of length to depth of the scour is approximately constant. The resulting expressions for y_s are:

$$\text{round jet: } \frac{y_s}{D_u} \sim \frac{U_o}{U_c} \sim \frac{U_o}{u_{*c}} \tag{6.21}$$

$$\text{plane jet: } \frac{y_s}{2B_u} \sim \left(\frac{U_o}{U_c}\right)^2 \sim \left(\frac{U_o}{u_{*c}}\right)^2 \tag{6.22}$$

where u_{*c} is the threshold shear velocity of the given bed material.

Original data have been replotted using these parameter groups in Figure 6.7. These dimensionless ratios were used to correlate experimental results and yield for a

$$\text{round jet: } \frac{y_s}{D_u} = 0.08 \frac{U_o}{u_{*c}} \tag{6.23}$$

$$\text{plane jet: } \frac{y_s}{2B_u} = 0.008 \left(\frac{U_o}{u_{*c}}\right)^2 \tag{6.24}$$

These expressions can be used to estimate scour depth. The lengths of the scour holes, L_s, are approximately 5 to 7 times the depths.

For *submerged vertical round jets*, the maximum velocity U_m along the sand bed after deflection of the jet depends upon $U_o D_u r^{-1}$. Therefore, if the width of the scour hole is taken to be equal to the value of r for which $U_m = U_c$, then the width and therefore the depth vary as:

$$\frac{y_s}{D_u} \sim \frac{U_o}{U_c} \sim \frac{U_o}{u_{*c}} \tag{6.25}$$

The original data for the static scour depth y_{ss} have been replotted for these parameter groups as shown in Figure 6.8, in which the static scour depth values have been used for y_s.

The rate of increase changes at $U_o/u_{*c} = 100$. For larger values of U_o/u_{*c} scour depth increases at a lower rate, most probably due to the effect of recirculating sand in suspension in the scour hole which dissipates energy and thereby decreases the erosive

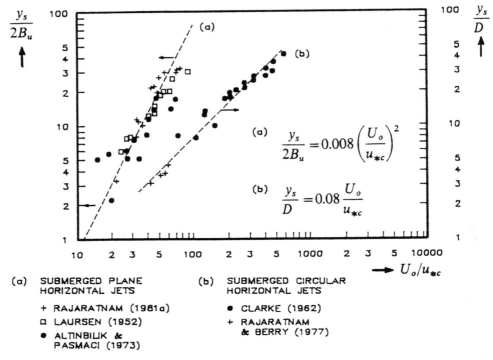

Figure 6.7. Maximum scour depth for submerged horizontal jets.; (a) plane, (b) circular.

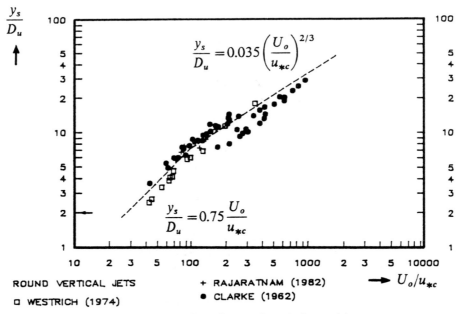

Figure 6.8. Maximum scour depth for submerged vertical round jets.

action of the jet. The distance H in the experiments with submerged nozzles does not appear to affect the results over a wide range of H/D_u values, but ultimately H must have an effect. For very large values of H, the local velocity is not large enough to produce scour (Figure. 6.6). The following equations of the straight lines on Figure 6.8 reflect the trends shown within a variation of $\pm 20\%$:

$$\frac{U_o}{u_{*c}} < 100 \qquad \frac{y_{ss}}{D_u} = 0.75 \frac{U_o}{u_{*c}} \tag{6.26a}$$

$$\frac{U_o}{u_{*c}} > 100 \qquad \frac{y_{ss}}{D_u} = 0.035 \left(\frac{U_o}{u_{*c}}\right)^{2/3} \tag{6.26b}$$

The average width/depth ratio for scour by vertical jets is 5.

For *submerged vertical plane jets* the available data by Rajaratnam and Akashi and Saito have been replotted in Figure 6.9 as $y_{ss}/2B_u$ versus U_o/u_{*c} using $H/2B_u$ as the third variable. There is some tendency of an increase of y_{ss} with H but the data do not justify a design relationship. However, the experimental results give some guidance for design.

6.6. SCOUR BELOW CULVERT OUTLETS

Culverts are probably the most numerous of all hydraulic structures and thus occur in many different forms throughout the world. Unprotected culvert outlets can induce

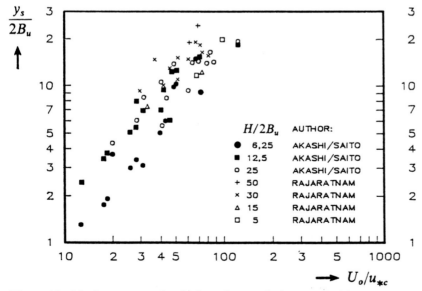

Figure 6.9. Maximum scour depths for submerged plane vertical jets.

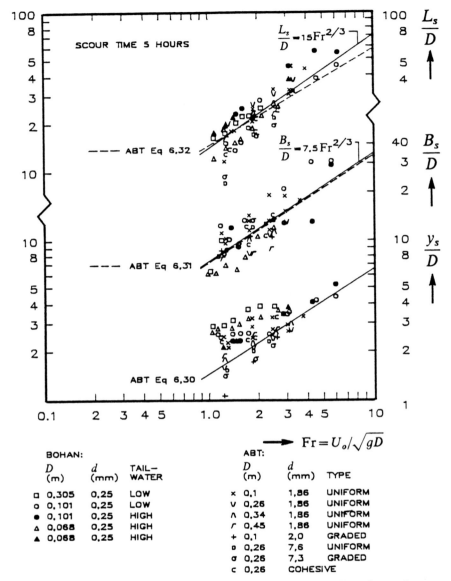

Figure 6.10. Dimensions of scour hole for culvert scour versus Froude number.

substantial scouring which can lead to undermining of the culvert and to embankment instability, as shown by Bohan (1970) among others. Bohan performed model tests on circular outlets under the following conditions:

— culvert size $D = 68, 101$ and 305 mm
— bed material size $d = 0.25$ mm

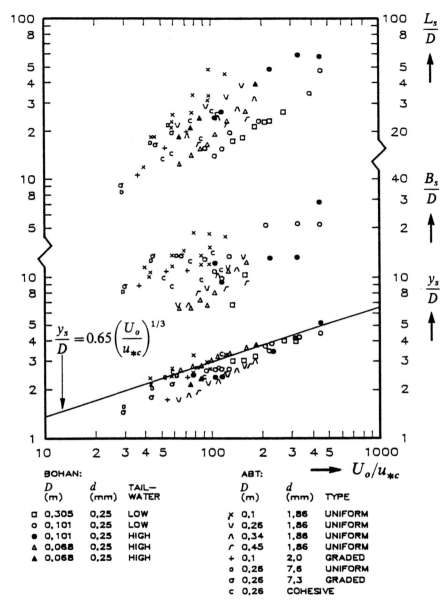

Figure 6.11. Dimensions of scour hole for culvert scour versus U_o/u_{*c}.

— variable tailwater depth: tailwater y_o was classified as "low" for $y_o/D < 0.5$ and as "high" for $y_o/D > 0.5$.

The experimental results were presented as dimensionless ratios using the culvert size as the length scale. His data are presented, using the Froude number $\mathrm{Fr} = U_o/\sqrt{gD}$

as suggested by Bohan, in Figure 6.10 and using U_o/u_{*c} in Figure 6.11; U_o is the mean velocity $Q/(\pi D^2/4)$, and u_{*c} is computed from the Shields' graph. The range of Froude numbers was from 1.06 to 6, so the pipe probably flowed "full" at the outlet end. No end wall was present so that scour could also develop *under* the culvert.

The size of the scour hole increased with time aproximately as $t^{0.1}$. The data in Figure 6.10 and 6.11 were extrapolated to $t=5$ hours with this relation for tests with durations shorter than 5 hours.

In order to protect the bed of the channel against scouring, the minimum stone size (stone density not specified) given by Bohan is:

$$\frac{d_s}{D} = 0.25 \text{Fr} \tag{6.27}$$

for "low" tailwater. For "high" tailwater he recommends slightly lower values (as mentioned, the dividing point for the tailwater depth is half the outflow diameter),

$$\frac{d_s}{D} = 0.25 \text{Fr} - 0.15 \tag{6.28}$$

Fletcher and Grace (1974) recommend a lining geometry as given in Figure 6.12. Simons and Stevens (1972) gave recommendations for the design of non-scouring and scouring bed protection, summarized in Figure 6.13 and 6.14. If some scour is allowed,

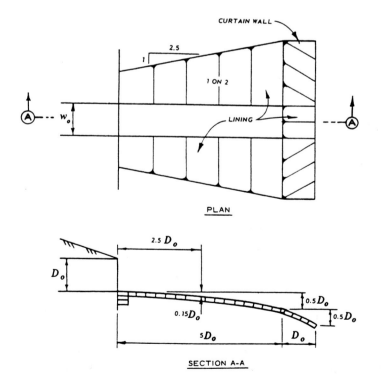

Figure 6.12. Recommended lined channel protection geometry downstream of a culvert (Fletcher and Grace 1974).

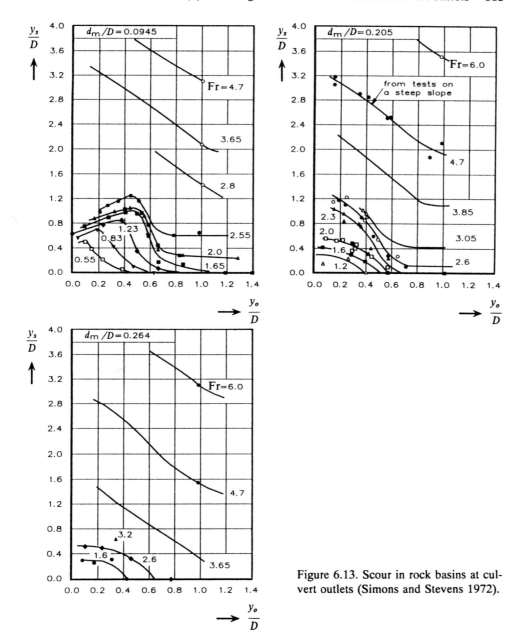

Figure 6.13. Scour in rock basins at culvert outlets (Simons and Stevens 1972).

bed protection thickness should be greater than the predicted scour depth. The allowance of some scour could give a more economical solution than using such large stones that any scour is prevented.

Ruff et al. (1982) reported results of systematic tests on culvert scour in flumes with widths of 1.2 and 6.0 m, using various materials, ranging from uniform sand

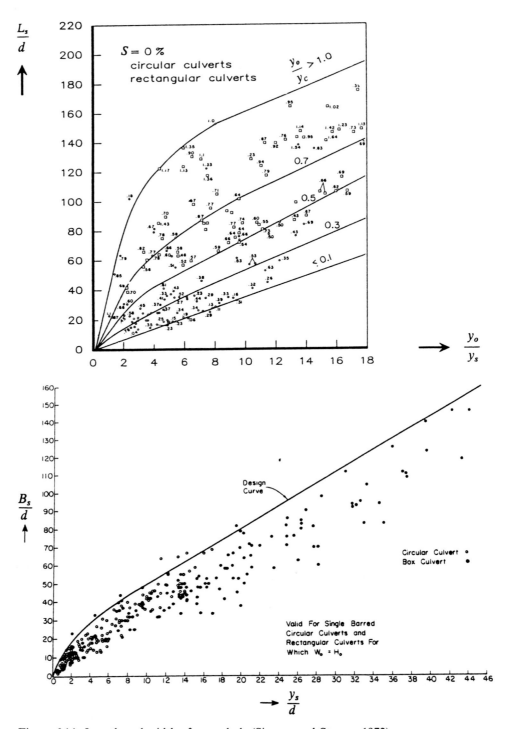

Figure 6.14. Length and width of scour hole (Simons and Stevens 1972).

($d = 1.86$ mm) to graded gravel ($d = 7.3$ mm) with pipe diameter D from 0.1 to 0.44 m. Little difference was obtained in the maximum scour hole dimensions as the tailwater varied from zero to $0.45D$. Their data have been replotted in Figures 6.10 and 6.11.

A few tests by Ruff et al. (1982) were conducted with an artificial cohesive material (58% sand, 15% silt, 27% clay) compacted to 90% of maximum density. No critical shear stress was given. The scour dimensions follow the general trend for sand and gravel if a value of $u_{*c} = 0.04$ m/s is used, an acceptable value for this type of material.

Abt et al. (1984) analysed the tests by Ruff et al. using their results for culvert diameters $D = 0.102$ and 0.254 m, tailwater depth $y_o = 0.45D$, and sand sizes 0.22 to 7.3 mm, both uniform and graded. Results were given in terms of

$$Q^* = \frac{Q}{\sqrt{gD^5}} = \frac{\pi}{4} \text{Fr} \tag{6.29}$$

Their correlations, based on experiments in the range $0.27 < \text{Fr} < 2.7$ and for $t = 5.3$ hours, are given in terms of Fr as follows:

$$\frac{y_s}{D} = 1.52 \text{Fr}^{0.63} \tag{6.30}$$

$$\frac{B_s}{D} = 7.44 \text{Fr}^{0.66} \tag{6.31}$$

$$\frac{L_s}{D} = 15.6 \text{Fr}^{0.58} \tag{6.32}$$

These equations are superimposed on Bohan's data in Figure 6.10 and indicate acceptable agreement except for Equation 6.30 which predicts too low values for the scour depth.

Introduction of sediment properties improved the coefficient of correlation for the scour depth from 0.72 to 0.83. The best correlation was obtained for the relationship

$$\frac{y_s}{D} = 3.18 \text{Fr}^{0.57} \left(\frac{d_{50}}{D}\right)^{0.114} \sigma_g^{-0.4} \tag{6.33}$$

where $\sigma_g = (d_{84}/d_{16})^{1/2}$. However, the effect of material gradation is not well defined, because three of five data sets were for essentially uniform material and two for σ_g equal to 4.38 and 4.78. A review of the original data shows that for graded gravel, scour depths are 15% less than for uniform gravel. The scour depth in the graded sand was 50% less for low Froude number and almost equal for large Froude number. Correlations for B_s and L_s were not improved by introducing the bed material properties.

Abt et al. (1985) also investigated the effect of culvert slope on scour. Slopes (up to 10%) increased dimensions of the scour hole. The scour depth was increased by as much as 15%, and the width and length increased by a maximum of 25%.

The effect of a vertical head wall at the end of the culvert on scour hole dimensions was not significant according to Mendoza et al. (1983). Near the head wall, the scour depth was almost equal to the maximum scour depth. If a head wall is placed below the culvert it should therefore extend to a depth at least equal to the maximum depth predicted for the scour hole.

6.6.1 Discussion of results on culvert scour

From the results on scour by culvert outlets by Bohan, Ruff et al. the following conclusions can be drawn:

- scour dimensions cannot be uniquely related to U_o/u_{*c} or exit Froude number due to the effect of the free surface on the scouring process (variable tail water)
- gradation has a limited effect on scour depth except at low exit velocities. Safe predictions can therefore be obtained using the median grain size d_{50}.
- some tests with a cohesive material show that scour dimensions will be of the same order as for sand and gravel. The value of u_{*c} to be used has to be obtained from tests with the cohesive material involved.
- scour depth is best related to U_o/u_{*c} whereas length and width are better related to the exit Froude number.

The information on cohesive materials is so limited that the relations for those materials are not adequate for design proposes.

The following design relations are suggested:

$$\text{scour depth} \quad \frac{y_s}{D} = 0.65 \left(\frac{U_o}{u_{*c}}\right)^{1/3} \qquad (6.34)$$

$$\text{scour width} \quad \frac{B_s}{D} = 7.5(\text{Fr})^{2/3} \qquad (6.35)$$

$$\text{scour length} \quad \frac{L_s}{D} = 15(\text{Fr})^{2/3} \qquad (6.36)$$

Equations 6.27 and 6.28 and the graphs by Simons and Stevens (1972) can be used to obtain design values for stable bed protections or safe rock basins.

6.7. SCOUR DUE TO PLUNGING JETS

6.7.1. Introduction

Excess water spilled from reservoirs is often conveyed via one of many spillway types to an energy dissipation structure or area. The structure may be a special stilling basin, or it may be the downstream river bed itself. If no stilling basin is provided, scour will occur due to the jet impinging on the river bed. The extent of the resulting scour

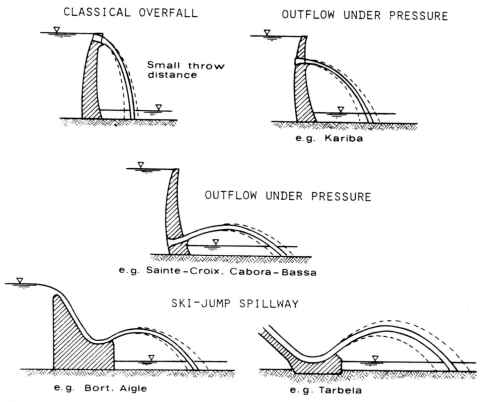

Figure 6.15. Jet types.

depends upon whether the bed consists of rock, cohesive or non-cohesive material. Some jet types are shown in Figure 6.15.

The erosion process is quite complex and depends upon the interaction of hydraulic and morphological factors. If the bed material consists of rock, scour will depend on rock type, weathering, the presence of fissures etc. For this type of material, experience is highly specialized and no general design relations can be given. For a review of available literature see Whittaker and Schleiss (1984), from which parts of this chapter were reproduced, and Häusler (1983). Available information on scour in non-cohesive material will be reviewed in Chapter 6.7.3. For field data, also cases with rock as bed material are included.

Scouring can have three major effects:

— the endangering of the stability of the structure itself by structural failure or increased seepage.
— the endangering of the stability of the downstream riverbed and side slopes.
— the formation of a mound of eroded material which can raise the tailwater level at the dam.

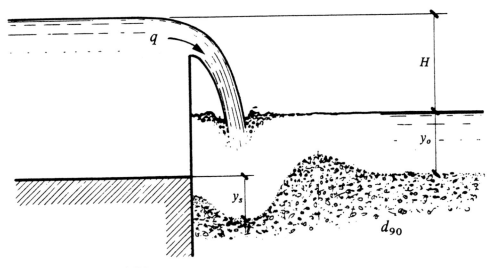

Figure 6.16. Free overfall jet scour.

6.7.2. Jet behaviour in a plunge pool

The general behaviour of a submerged jet is presented in Chapter 6.2. The relations given for jet dimension and velocity can also be used for plunging jets if jet velocity and dimension at impact are taken as initial values. Jet velocity varies with s^{-1} for round jets and with $s^{-1/2}$ for plane jets, where s is the axial distance in the plunge basin from the point of impact.

For practical purposes, the depth of jet penetration may be taken as $20D_u$ for round jets and $80B_u$ for plane jets according to Häusler (1983), where D_u and $2B_u$ represent the jet size at impact. However, these values can vary with the initial jet velocity at the point of impact and the resistance of the river bed against erosion.

Aeration of the jet during its flight in air can also affect the jet velocity, but Häusler (1983) assumed that for conventional design values, the core region will persist throughout the drop to the tailwater level. He recommends either to ignore aeration or to consider its effect through an appropriate reduction of the jet width at the point of impact.

6.7.3. Scour by plunging jets

Data on scour have generally been obtained from small scale tests with non-cohesive material or from field observations for which the material properties are unspecified. A selection of available relations shows a variety of forms. Veronese (1937) presented a relationship for plane plunging jets in a flume with $B = 0.5$ m, discharge $q = 0.01$ to $0.07 \, \text{m}^2/\text{s}$ and grain sizes $\bar{d} = 9, 14, 21$ and 36 mm:

$$y_s + y_o = 3.68 H^{0.225} q^{0.54} \bar{d}^{-0.42} \tag{6.37}$$

where $\bar{d}$ is expressed in mm and H is the difference in upstream and downstream water-level (Figure 6.16). Veronese found from a second series of tests that scour varied less than predicted by Equation 6.37 for $\bar{d} < 5$ mm. The scour is then given by:

$$y_s + y_o = 1.9 H^{0.225} q^{0.54} \tag{6.38}$$

This relation is suggested by USBR (1973) as a limiting scour depth.

Eggenberger (1944) found for an overflow weir type, expressing d_{90} in mm:

$$y_s + y_o = 22.88 H^{0.5} q^{0.6} d_{90}^{0.40} \tag{6.39}$$

Equation 6.39, predicts very high values of $(y_s + y_o)$, which are probably too high in view of the experimental procedure (Chapter 7) of removing part of the bed material to accelerate the scouring process.

Damle et al. (1966) evaluated model data and some field data for Indian dams with ski jumps and gave as a best-fit relation in metric units:

$$y_s + y_o = 0.55 (qH)^{0.5} \tag{6.40}$$

Chian Min Wu (1973) used model and prototype data from dams in Taiwan and found

$$y_s + y_o = 1.18 q^{0.51} H^{0.235} \tag{6.41}$$

Martins (1975) derived an empirical relation from some prototype observations:

$$y_s + y_o = 1.5 q^{0.6} Z_2^{0.1} \tag{6.42}$$

where Z_2 is the difference in elevation between the free surface of the reservoir and the lip of the flip bucket.

Mason (1984) and Mason and Arumugam (1985) analysed model and prototype data and gave the following relations:

Model data

$$y_s + y_o = 3.27 q^{0.6} H^{0.05} y_o^{0.15} g^{-0.3} d_m^{-0.1} \tag{6.43}$$

Prototype data (and model data)

$$y_s + y_o = (6.42 - 3.1 H^{0.1}) q^{(0.6 - 0.0033 H)} H^{(0.05 + 0.005 H)} y_o^{0.15} g^{-0.3} d_m^{-0.1} \tag{6.44}$$

The value of d_m was assumed to be 0.25 m for prototype data (d_m is the mean grain size given in meters for these two equations). The relation for model data only is dimensionally correct and satisfies Froude's scaling law. The coefficient of variation was 25% for both model data (47 cases) and for prototype data (26 cases). Outlet types included free overfalls, low level outlets, spillway chute flip buckets and tunnel outlets.

Discussion

Results from the analysis of model and prototype data show great variations in the form of equations and coefficients. Exponents of q vary only from 0.5 to 0.7 whereas the exponents of H vary from 0.05 to 0.5.

A simple theoretical analysis, assuming a plane jet at impact with the downstream water level and a scour depth up to a level for which the maximum jet velocity is equal to some critical velocity for bed material erosion, $U_c \sim (\Delta g d)^{0.5}$, leads to an expression of the form

$$(y_s + y_o) \sim q H^{0.5} d^{-0.5} \qquad (6.45)$$

The exponents are much greater than those from experimental data. Apparently other mechanisms of energy dissipation play a role such as those due to the presence of eroded material in the scour hole.

Whittaker and Schleiss (1984) made a comparison of the various relations for a practical case, the Cabora-Bassa dam in Mozambique. This dam has a middle-level outlet (Figure 6.17).

The maximum discharge through 8 sluices is 13,100 m³/s for a reservoir level of 325 m, and the downstream water level is 225 m. The elevation of the lip of the spillway sluices is 244 m. The value of q is estimated to be 275 m²/s, the downstream water depth is about $y_o = 40$ m.

In the model tests, performed at a scale of 1:75, the movable bed was composed of gravel with $d_{85} = 35$ mm, $d_{50} = 28$ mm and $d_{15} = 13$ mm. The bed was weakly aggregated with aluminous cement. The corresponding prototype sizes are estimated approximately as $d_{85} = 2.6$ m, $d_{50} = 2.1$ m and $d_{15} = 1.0$ m. The modelled scour depth for the maximum discharge was 75 m (Quintella and Da Cruz 1982). In February 1982 $(y_s + y_o)$ was measured to be approximately 68 m. The values of scour depth predicted by means of the various formulae are listed in Table 6.1.

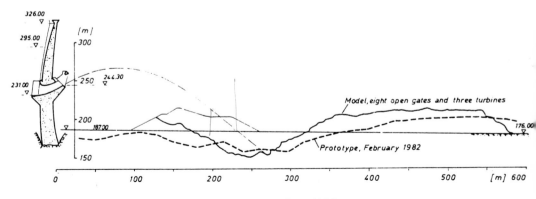

Figure 6.17. Cabora-Bassa Dam (Quintella and Da Cruz 1983).

Table 6.1 Scour predictions Cabora-Bassa dam.

Formula	Equation	Predicted scour depth (m)
Veronese*	6.37	52
Damle	6.40	91
Chian Min Wu	6.41	61
Martins ($Z_2 = 82$ m)	6.42	68
Mason (model)*	6.43	98
Mason (proto)	6.44	71
Same for $d = 0.25$ m	6.44	87

*The equations of Veronese and Mason (model) have been used at a model scale of 1:100 to obtain values of q and H in the range of most model experiments. Predicted scour depth has then been translated to prototype scale by multiplying with a factor 100.

Most relations predict a value in the right order of magnitude, although differences between various relations are large, as might be expected.

In conclusion: the available relations give an indication of scour to be expected in coarse non-cohesive material or fissured rock. For scour problems related to the construction of high-head dams, detailed studies including model studies, should be performed for each case.

REFERENCES

Abt, S.R. et al. 1984. Unified culvert scour determination. *J. Hydr. Eng., ASCE* 110(10); 1475–1479.

Abt, S.R. et al. 1985. Culvert slope effects on outlet scour. *J. Hydr. Eng., ASCE* 111(10); 1363–1367.

Akashi, N. & T. Saito 1984. Estimation of equilibrium scour depth from submerged inpinged jet. *Proc. 4th Congress Asian and Pacific Division IAHR, Chiang Mai, Thailand*; 167–181.

Altinbilik, H.D. & Y. Basmaci 1973. Localized scour downstream of outlet structures. *Proc. 11th Congress on large dams, Madrid* II; 105–121.

Bohan, J.P. 1970. *Erosion and rip rap requirements at culvert and storm-drain outlets*. U.S. Army Eng. Waterways Exp. St. Vicksburg Res. Rep., H-70-2.

Chian Min Wu 1973. Scour at downstream end of dams in Taiwan. *Proc. IAHR Symp. on river mechanics, Bangkok*, paper A13.

Clarke, F.R.W. 1962. The action of submerged jets on moveable material. Ph.D. Thesis, Imperial College, London.

Damle, P.M. et al. 1966. Evaluation of scour below ski-jump buckets of spillways. *Proc. Golden Jubilee Symp., Poona*; 154–164.

Doddiah, D. et al. 1953. Scour from jets. *Proc. IAHR/ASCE Conference, Minnesota*; 161–169.

Eggenberger, W. 1944. *Die Kolkbildung beim reinen Überströmen und bei der Kombination Überströmen-Unterströmen*. Mitt. Versuchsanstalt für Wasserbau E.T.H. Zurich, Nr. 5.

Fletcher, B.P. & J.L. Grace Jr. 1974. *Practical guidance for design of lined channel expansions at culvert outlets*. U.S. Army Eng. Waterways Exp. St. Vicksburg Techn. Rep., H-74-9.

Häusler, E. 1983. Spillways and outlets with high energy concentration. *Trans. Int. Symp. on the lay-out of dams in narrow gorges, ICOLD, Rio de Janeiro II*; 77–194.

Johnson, G. 1967. The effect of entrained air on the scouring capacity of water jets. *Proc. 12th IAHR Congress, Ft. Collins*, 3, paper C26; 218–226.

Kobus, H. et al. 1979. Flow field and scouring effects of steady and pulsating jets impinging on a movable bed. *J. Hydr. Res.* 17(3); 175–192.

Laursen, E.M. 1952. Observations on the nature of scour. *Proc. 5th Hydr. Conf. Univ. of Iowa, Studies in Engineering. Bull.* 34; 179–197.

Martins, R.B.F. 1975. Scouring of rocky river beds by free jet spillways. *Water Power and Dam Constructions*, April; 152–153.

Mason, P.J. 1984. Erosion of plunge pools downstream of dams due to the action of free-trajectory jets. *Proc. Instn. Civ. Engrs.* Part 1, 76; 523–537.

Mason, P.J. & K. Arumugam 1985. Free jet scour below dams and flip buckets. *J. Hydr. Eng., ASCE* 111(2); 220–235.

Mendoza, C. et al. 1983. Headwall influence on scour at culvert outlets. *J. Hydr. Eng., ASCE* 109(7); 1056–1060.

Quintella, A.C. & A.A. Da Cruz 1983. Cabora-Bassa Dam Spillway. Conception, hydraulic model studies and prototype behaviour. *Trans. Int. Symp. on the lay-out of dams in narrow gorges, ICOLD, Rio de Janeiro*.

Rajaratnam, N. 1976. *Turbulent jets*. Elsevier, Amsterdam.

Rajaratnam, N. 1981a. Erosion by plane turbulent jets. *J. Hydr. Res.* 19(4); 339–358.

Rajaratnam, N. 1981b. *Further studies on the erosion of sand beds by plane water jets. Study 1: Erosion of sand beds by obliquely impinging plane turbulent submerged water jets*. Univ. of Alberta, Dept. of Civil Engineering, Report WRE 81, Edmonton, Alberta.

Rajaratnam, N. 1981c. *Erosion of sand beds by circular impinging jets with minimum tailwater*. Univ. of Alberta, Dept. of Civil Engineering, Edmonton, Alberta.

Rajaratnam, N. 1982a. Erosion by submerged circular jets. *Proc. ASCE* 108(HY2); 262–267.

Rajaratnam, N. 1982b. Erosion by unsubmerged plane water jets. In: *Applying research to hydraulic practice, Jackson, 1982*. ASCE New York; 280–288.

Rajaratnam, N. & B. Berry 1977. Erosion by circular turbulent wall jets. *J. Hydr. Res.* 15(3); 277–289.

Rajaratnam, N. & R.K. MacDougall 1983. Erosion by plane wall jets with minimum tail water. *J. Hydr. Eng., ASCE* 109(7); 1061–1064.

Ruff, J.F. et al. 1982. *Scour at culvert outlets in mixed bed materials*. Federal Highway Adm. USA, Rep. FHWA/RD-82/011.

Simons, D.B. & M.A. Stevens 1972. *River Mechanics*, Vol. II, H.W. Shen, (ed.). Ch. 24. Scour control in rock basins at culvert outlets.

Tarapore, Z.S. 1956. Scour below a submerged sluice gate. M. Sc. Thesis, Univ. of Minnesota, Minneapolis.

USBR 1973. *Design of small dams* (2nd ed.). Water Resources Techn. Publ., Denver.

Vanoni, V.A. (ed.) 1975. *Sedimentation Engineering*. ASCE, New York.

Veronese, V. 1937. Erosion de fond en aval d'une décharge. *IAHR Meeting for hydraulic works*, Berlin.

Westrich, B. & H. Kobus 1973. Erosion of a uniform sand by continuous and pulsating jets. *Proc. 15th IAHR Congress, Istanbul*, 1; 91–98.

Westrich, B. 1974. Erosion eines gleichkörnigen Sandbettes durch stationäre und pulsierende Strahlen. Diss. T.U. Karlsruhe.

Whittaker, J.G. 1984. Time development and local scour by jets (private communication).

Whittaker J.G. & A. Schleiss 1984. *Scour related to energy dissipators for high head structures*. Mitt. Nr. 73 VAW/ETH, Zurich.

CHAPTER 7

Scour below low head structures

H.N.C. BREUSERS

7.1. INTRODUCTION

Information on scour due to flows at lower heads and velocities includes an extension of Chapter 6 to comprise flows over and under gates, weirs and low dams. These are mostly somewhat idealized laboratory studies at small scale or project related model studies which are of little general significance. The experiments have also been limited to studies using coarse sediments with an emphasis on equilibrium scour. A lack of verification by field data limits the usefulness of the results.

A different type of study is the extensive work done in connection with the Netherlands Delta Project and parallel work in Germany. These include studies of the time dependence of the rates of scour which are characteristic of the fine sediments found in deltas.

7.2. WEIRS AND DAMS ON COARSE SEDIMENTS; STILLING BASINS

Weirs and low dams are frequently installed to control the water level in canals and rivers. Flow under or over such structures has a considerable potential for scour even for comparatively low heads. Figure 7.1 shows the various types of flow and the principal dimensions.

$q = q_o + q_u$ (m²/s)
q_o = flow over a weir or dam (m²/s)
q_u = flow under a weir or gate (m²/s)
y_o = downstream water depth (m)
y_s = depth of scour below original bed level (m)
L = length of fixed (protected) bed (m)
$H = (H_1 - y_o)$, as shown (m)

Because scour in coarse sediment develops very rapidly, only the equilibrium scour depth has been reported in most cases. Kotoulas (1967) performed a rather extensive set of tests and found that about 64% of the final scour occurred in the first 20 seconds, and about 97% of the scour depth was attained in just 2 hours.

In tests with graded sediments, various observers reported segregation of the bed

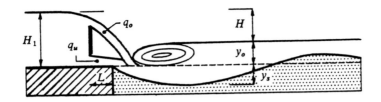

Figure 7.1. Definition sketch.

material in the scour hole. As a consequence, Hartung (1957) and Kotoulas (1967) used d_{85} and d_{95}, respectively, to characterize the bed material. In the formulas the grain size d is expressed in mm unless otherwise specified.

Schoklitsch (1932, 1935) evaluated separately model tests for (a) overflow alone ($q_u = 0$ and $L = 0$),

$$y_s + y_o = 4.75 H^{0.2} q^{0.57} d_{90}^{0.32} \tag{7.1}$$

and for (b) underflow alone with a short horizontal sill ($L = 1.5H$),

$$y_s = 0.378 H_1^{0.5} q^{0.35} + 2.15 a \tag{7.2}$$

where a is the level of the downstream riverbed *below* the sill level. The effect of grain size on the scour was negligible for the range $d = 1.5$ to 12 mm in the model tests; the difference H_1 between upstream water level and sill level (Figure 7.1) varied from 0.3 to 1.0 m. The tail water depth was not varied independently.

Schoklitsch (1949) also discusses the effect of sill shape and form of the hydraulic structure, but most of the information is restricted to the specific shapes used.

Veronese (1937), (also discussed in Chapter 6.7.3), found for plunging jets

$$y_s + y_o = 3.68 H^{0.225} q^{0.54} \bar{d}^{-0.42} \tag{7.3}$$

for

$\bar{d} = 9, 14, 21, 36$ mm
$q = 0.01 - 0.07$ m^2/s

Jaeger (1939) reanalysed the data obtained by Veronese and gave the relationship

$$y_s + y_o = 6 H^{0.25} q^{0.5} (y_o/d_{90})^{0.33} \tag{7.4}$$

a dimensionally consistent formula.

Eggenberger (1944) conducted tests in a laboratory flume and proposed the relationship

$$y_s + y_o = 22.9 H^{0.5} q^{0.6} d_{90}^{-0.4} \tag{7.5}$$

for the conditions:

$d_{90} = 1.2, 3.5, 7.5$ mm
$q_o\ = 0.006 - 0.024$ m^2
$H\ = 0.19 - 0.35$ m

Eggenberger as well as Müller (1944) shortened their tests by removing material systematically from the downstream side of the scour hole until equilibrium was assumed to have been reached. This procedure probably led to overestimates of scour depth because the scour reducing action of the recirculation of eroded material within the scour hole was diminished and because the downstream barrier was lowered. The presumption is that the mound downstream of the scour hole will erode away, something that does not always occur.

Müller (1944) performed tests (a) for underflow only, and (b) for combined underflow and overflow. For the underflow case, Müller reported two patterns of flow with different quantities of scour which he describes as (a) a submerged wavy jet and (b) a free wavy jet. For the two cases the coefficients in Equation 7.5 were (a) 10.35 and (b) 15.4 respectively.

The test conditions were as follows:

flume width $B = 0.9$ m

$d_{50} = 0.43, 0.76, 1.9$ and 3.67 mm

$q = 0.004-0.025$ m²/s

$H = 0.50-0.12$ m

$L = 0.06$ m

For combined underflow and overflow, Eggenberger obtained the following constants in Equation 7.5 for what he described as predominantly plunging jets:

q_o/q_u	coefficient in Equation 7.5
2	15.6
3	18.5
4	20.4
∞	22.9

Hartung (1957) repeated the tests of Veronese, using $q = 0.0036-0.21$ m³s, grainsize from 2 to 15 mm (5 grades) and obtained:

$$y_s + y_o = 12.4 H^{0.36} q^{0.64} d_{85}^{-0.32} \tag{7.6}$$

Scour depths y_s, reported by the authors other than Eggenberger and Müller were generally less, principally because of Eggenberger's and Müller's experimental technique. Kotoulas (1970) remarks that values obtained from the Eggenberger and Müller relations are usually reduced with some 30 to 50% by designers in engineering practice.

Shalash (1959) studied the influence of apron length L for underflow alone. The ranges for his tests were $q = 0.011-0.027$ m²/s and $d_{50} = 0.7, 1.9$ and 2.65 mm. For an apron with a horizontal end sill:

$$y_s + y_o = 9.65 H^{0.5} q^{0.6} d_{90}^{-0.4} (L_{min}/L)^{0.6} \tag{7.7}$$

in which L = length of apron, and $L_{min} = 1.5H$.

An idea of the reliability of the relations can be obtained only by comparing the values predicted for specific examples. For example, for $H=5$ m, $q=4$ m^2/s, $y_o=4$ m and $d_{90}=50$ mm, the various equations predict the following values for an overspill:

Author	Equation	$y_s + y_o$	y_s
Schoklitsch	(7.1)	9.5 m	5.5 m
Veronese	(7.3)	7.5 m	3.5 m
Jaeger	(7.4)	7.7 m	3.7 m
Eggenberger	(7.5)	24.5 m	20.5 m
Hartung	(7.6)	15.3 m	11.3 m

The equations of Schoklitsch and Veronese, being dimensional, were first applied at an assumed model length scale of 1:25 and the computed scour depth was then converted to prototype dimensions using the same length scale.

Even without Eggenberger's formula, the range of predicted depths is still large and clearly unsatisfactory for field situations. Correspondingly, the equations can only be used for first estimates, and only if the ranges of values of the dimensionless variables, such as H/d, etc. coincide with those used in the foregoing tests. Any final prediction must rest on specific model tests.

If the same values are used for the underflow situation, the computed depths are:

Author	Equation	$y_s + y_o$	y_s
Schoklitsch, $a=0$, $H_1=9$ m	(7.4)	5.7 m	1.7 m
Müller, submerged jet	(7.5)	11.1 m	7.1 m
Shalash, $L=L_{min}$	(7.7)	10.4 m	6.4 m

also too wide a range for prediction to be useful.

Effect of stilling basins. Stilling basins are sometimes used to protect the region of intensive scour. However, little generally applicable information is available on the dimensions of the scour downstream of a stilling basin. Novak (1955, 1961) stated that use of a stilling basin which is sufficiently long to contain the hydraulic jump will reduce the scour to some 45 to 65% of that without a stilling basin. Scour predicted by Jaeger (Equation 7.4) was used as the reference for scour without a stilling basin. The lowest value occurs for $y_o/y_{o\,min}=1.6$ whereas the highest is for $y_o/y_{o\,min}=1$, where $y_{o\,min}$ is the minimum downstream water depth necessary to form a hydraulic jump. Çatakli et al. (1973) found from small-scale experiments on a spillway having a stilling basin with length $5y_o$:

$$y_s + y_o = 1.6 H_1^{0.2} q^{0.6} d_{90}^{-0.1} \tag{7.8}$$

$q = 0.05$ to 0.1 m^2/s

$d_{90} = 1$ and 10 mm

For the example given above a value of $y_s + y_o = 3.85$ m is predicted, which is unrealistically small in view of the fact that $y_o = 4.0$ m.

Information on scour downstream from dams of the type shown in Figure 7.1 is clearly unsatisfactory for use in design. Several of the equations cannot readily be extended to prototype scale. Even those which are dimensionally consistent lack corroboration from full scale tests, and the variability of the predictions point at considerable inconsistency even at model scale. In actual cases, three-dimensional effects due to horizontal expansions can increase scour. Thus the results presented must be considered as only rough estimates.

Specific designs require model tests to simulate specific geometric features and to provide comparable conditions of scour.

7.3. TIME DEPENDENT SCOUR BELOW SILLS AND WEIRS IN FINE SEDIMENT

The information presented in Section 7.2 is not applicable to structures in deltaic areas which are characterized by fine sediments, particularly when the time factor is important, as in the case of closure of estuary branches.

The Netherlands Delta Project, for example, required special investigations because of risk of failure of the structure due to scour, aggravated by the possibility of liquefaction of loosely packed sand layers which become disturbed by the scouring and can lead to undermining.

Experiments generally show (Li 1955, Ghetti and Zanovello 1954, Breusers 1966, Dietz 1969) that for a given geometric arrangement, the shape of the scour hole is almost independent of flow velocity and bed material if compared at equivalent values of $y_{s\,max}/y_o$. This increases the applicability of small-scale tests as long as the time scale of the scouring process is known. The scour geometry is independent of grain size and velocity only if u_* is large relative to u_{*c} so that all grains are easily moved. The depth of scour in a non-uniform sediment (large σ_g) is usually much less than in uniform sediment because of armouring of the bed. It should be noted that most experiments have been made with uniform sands.

The flow field in a two-dimensional scour hole is generally similar to that of a highly turbulent mixing layer. A weak return current is formed near the bed in the first part of the scour hole, this ends at the point of maximum scour depth. In this region the flow is very turbulent, and large vortices intermittently erode and transport bed material. Downstream of the deepest point, the flow accelerates, velocity profiles slowly return to normal and turbulence decreases (Breusers 1966, Raudkivi 1963, 1967), see Figure 7.2.

In three-dimensional situations, for example scour in a horizontal contraction, vortices with vertical-axis are important because of their intensive action and erosive capacity. Scour depths in three-dimensional cases generally tend to be much greater than in comparable two-dimensional cases. It is important to avoid strong horizontal velocity gradients which generate vortices with vertical axis.

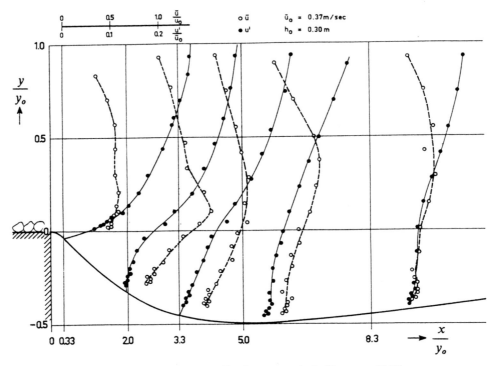

Figure 7.2. Velocity and turbulence profiles in a scour hole (Breusers 1966).

Literature review

Li (1955) investigated scour below a submerged sluice gate and Ghetti and Zanovello (1954) did the same for a spillway. Both found the scour hole shape to be independent of flow velocity and bed material.

Breusers (1966, 1967) reported systematic studies on two-dimensional scour. The impetus for these studies was provided by the Netherlands Delta Project where sluices and dams had to be built on loosely packed sediments. Observations on scour with various bed materials, flow velocities, and situation geometries showed that for a given geometry, similarity of scour hole shape was satisfied by the relationship (see Figure 7.3):

$$\frac{y_s(x, t)}{y_o} = f\left(\frac{t}{t_1}\right) \qquad (7.9)$$

where t_1 is the time at which $y_{s\,max} = y_o$. For definitions see Figure 7.2. Bed materials used were sand, bakelite ($\rho_s = 1350$ kg/m^3) and polystyrene ($\rho_s = 1045$ kg/m^3). The maximum scour depth as a function of time was:

$$\frac{y_{s\,max}}{y_o} = \left(\frac{t}{t_1}\right)^{0.38} \qquad (7.10)$$

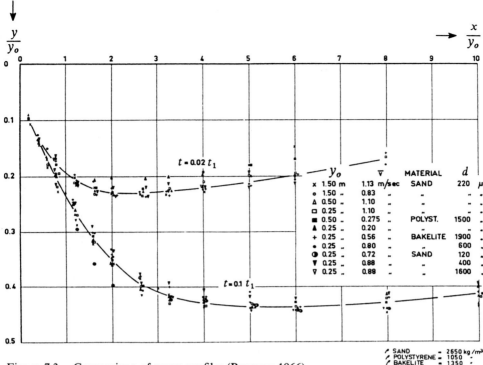

Figure 7.3. Comparison of scour profiles (Breusers 1966).

for all two-dimensional tests. The tests data (Figure 7.4) support the concept that scour continues to increase with time.

The characteristic time t_1 was expressed, on the basis of more than 250 tests, as:

$$t_1 = 330 \Delta^{1.7} y_o^2 (\alpha \bar{U} - \bar{U}_c)^{-4.3} \tag{7.11}$$

in which:

t_1 = time in hours at which $y_{s\,max} = y_o$
Δ = $(\rho_s - \rho_w)/(\rho_w)$
α = factor dependent on flow geometry
U = average velocity at end of bottom protection ($x=0$)
U_c = critical mean velocity evaluated from Shields' curve

Some results for scour downstream of a low sill and a fixed bed protection with roughness are presented in Figure 7.5. The scour is seen to increase with sill height and to decrease with increasing fixed bed roughness. For predicting scour downstream of a rough rip-rap type bed protection, α was related to the average relative turbulence intensity $\overline{u'}/U$ at $x=0$ by

$$\alpha = 1 + 3(\overline{u'}/U) \tag{7.12}$$

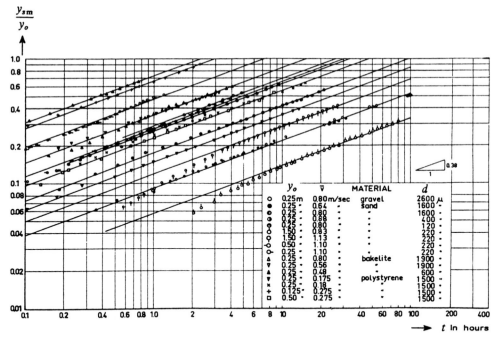

Figure 7.4. Increase in maximum depth of scour ($y_{s\max}$) for scouring downstream of a rough bed.

Table 7.1. Values of the factor α.

$\dfrac{y_D}{y_o}$	$\dfrac{L}{y_o}$	α Bed protection Smooth	α Rough
0	10	2.0	1.5
0.3	1 to 15	2.5	2.0
0.6	3	3.2	3.0
0.6	10	2.9	2.5

where $\overline{u'}$ is the depth-averaged root mean square value of the turbulent velocity fluctuation in the x-direction.

Equation 7.12 is also valid for a low sill. For higher sills or a smooth bed protection (e.g. a concrete apron) where distortion of the velocity profile occurs a correction factor α_u must be employed:

$$\alpha = \alpha_u(1 + 3\overline{u'}/U) \tag{7.13}$$

Indicative values for α, including effects of turbulence *and* velocity profile are listed in Table 7.1 (Delft Hydraulics Laboratory 1972a). Symbols used in Table 7.1 are defined in Figure 7.6.

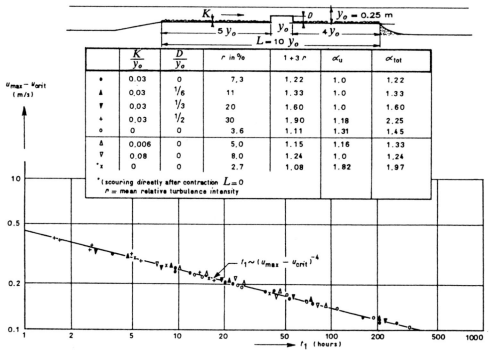

Figure 7.5. Influence of flow conditions on time t_1.

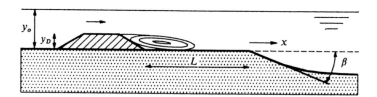

Figure 7.6. Definition sketch.

Note: "rough" was taken to be roughness/depth ratios in excess of 0.02. A rough bed protection reduces the rate of scouring by modifying the velocity distribution at $x=0$. Higher sills cause greater turbulence, greater distortion of the velocity profile, and consequently larger values of α.

Raudkivi (1967) reported measurements of the velocity distribution and bed shear stress in a scour hole (Figure 7.7). Bed shear stresses were found to be much less than the critical values for the beginning of transport (evaluated from Shields), which indicates that the scouring process is strongly influenced by turbulence.

Dietz (1969) performed extensive research on two-dimensional scour downstream from horizontal beds and low sills. He used bed material of sand, lignite ($\rho_s = 1370$ kg/m^3) and polystyrol ($\rho_s = 1045$ kg/m^3), and y_o varied from 0.125 m to 0.25 m. His results generally agreed with those of Breusers (1966, 1967). The value of t_1 was described by

$$t_{1(\text{hours})} = 48 \, y_o^{1.75} \Delta^{1.5} (U_{\max} - \bar{U}_c)^{-4} \tag{7.14}$$

where $U_{\max} = \alpha U$

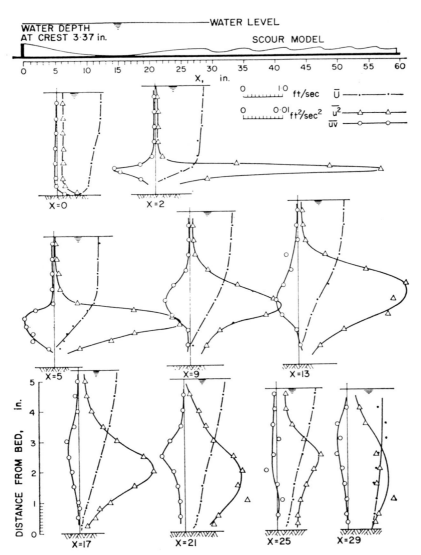

Figure 7.7. Scour model: profiles of the mean and fluctuating longitudinal velocities and the turbulent shear (Raudkivi 1967).

Note: An increase of the exponent of y_o is coupled with an increase in the value of the constant in Equation 7.14. Consequently the difference between Equations 7.14 and 7.11 is less than appears to be the case.

Dietz observed a tendency towards equilibrium in some of his tests. A tentative expression for the limiting scour depth is:

$$\frac{y_{s\,max(equil)}}{y_o} = \frac{U_{max} - U_c}{U_c} \qquad (7.15)$$

Equation 7.15 predicts a very large equilibrium scour depth. For example, with $\alpha = 2$ and $U = 3U_c$ the equilibrium scour depth is $5y_o$.

Dietz also investigated the slope of the initial part of the scour hole (β) which is of importance for the stability of adjacent structures or bed protection. Cotg β was a function of $(U - U_c)D_*/w$, where $D_* = d(\Delta g/\nu^2)^{1/3}$. The experimental results are shown in Figure 7.8.

The angle β decreased with decreasing grain size d. These results run counter to findings of the Delft Hydraulics Laboratory (1972b), in which β was observed to be essentially independent of grain size, except for a fine sand (120 μm) which yielded a generally steeper slope. Some indicative values of β for two-dimensional situations (Delft Hydraulics Laboratory 1972a) are listed in Table 7.2 (symbols defined in Figure 7.6). A smooth bed protection results in a steeper initial slope because the water

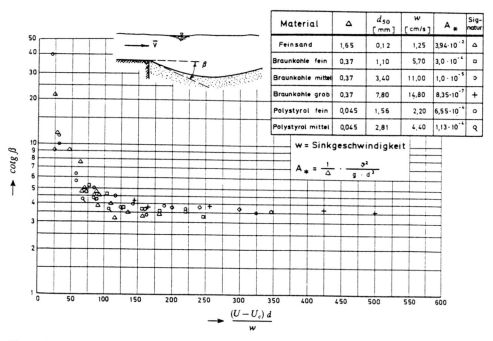

Figure 7.8. Initial slope of scour hole (Dietz 1969).

Table 7.2. Indicative values of β, with reference to Figure 7.6.

$\dfrac{y_D}{y_o}$	$\dfrac{L}{y_o}$	cotg β Bed protection Smooth	cotg β Rough
0	10	4	6
0.3	1	4	7
0.3	5	3	3.5
0.3	15	3.5	4.5
0.6	3	3	4.5
0.6	7	2.5	3
0.6	12	2.5	3.5

particles near the bed have more momentum causing a more rapid expansion of the flow in the scour hole. If only a very short section downstream of the sill is protected, the strong vortex generated by the sill structure transports material back towards the sill and the slope is flatter. Intermediate values of L/y_o are associated with steeper slopes because the protection ends approximately where the vortex ends, so that large scale turbulence attacks the first part of the scour hole. As shown in the following paragraphs, slopes in three-dimensional situations are generally much steeper.

The foregoing discussion involved patterns of two-dimensional flows only, though many field situations are three-dimensional, at least to some degree. A three-dimensional pattern was created in a laboratory flume for scour downstream of a partial channel constriction (Delft Hydraulics Laboratory 1972b, c; Van der Meulen and Vinjé 1975). The flumes were wide ($B = 10$, 5 and 2.5 m) and the constriction was a thin vertical board of length $0.1B$ at one side of the channel (Figure 7.9). The board was placed on a sill whose height was varied in the range $y_D/y_o = 0.6$, 0.3 and 0. The length of protected bed was maintained at $10 y_o$. In all, 110 tests were performed to determine the scour for these conditions which represents situations occurring in closures of estuarine channels. The main conclusions were that the shape of the scour hole is independent of bed material and flow velocity, and that Equation 7.11 is equally applicable to three-dimensional situations, if α is assigned greater values. However, a different function was required for the prediction of $y_{s\,max}$. For small scour depths an exponential function of the form of Equation 7.10 gave a good fit; the exponent took on values > 0.38. In the later stages of scour development a logarithmic function was found to be more appropriate. Some of the data from the study are presented in Figure 7.10.

Some indicative values of α for $L/y_o = 10$ are shown in Table 7.3. Table 7.3 shows that α increases strongly with sill height, this demonstrates that scour increases due to the decelerated flow downstream of the sill, which promotes the contraction effect. A typical scour hole is illustrated in Fig. 7.11, where scoured depths are given in m for the prototype situation. For evacuation sluices and closure works with $L/y_o = 10$, typical values of α lie in the range $2.5 < \alpha < 4$ for three-dimensional situations. The value of α can be decreased if care is taken in the design to minimize or avoid strong contractions and horizontal velocity gradients.

Scour hole slopes measured along a flow line through the point of maximum scour depth are generally steeper for the three-dimensional cases than for the two-dimensional ones due to the erosive action of vortices with vertical axes. A tentative relation (de Graauw, 1983) is

$$\cot \beta = 2.3 + (\alpha - 1.3)^{-1} \quad \text{for } \alpha > 1.5 \tag{7.16}$$

The values of α and β may be reduced by either lengthening the bed protection or making the bed rougher. Optimum relative roughness (k/y_o) is 0.025 to 0.05. The value of α can be predicted from the relationship (de Graauw and Pilarczyk 1981):

$$\alpha(L/y_o) = 1.5 + (1.75\alpha_{10} - 2.35) \exp(-0.045 \, L/y_o) \tag{7.17}$$

where $\alpha_{10} = \alpha$ at $L/y_o = 10$.

This relationship is valid for $L/y_o > 5$. The end of the bed protection should be flexible to prevent failure due to undermining, and should contain enough stone or other protection material to remain intact if deformation does occur. Moreover, in tidal flow zones the protective ballast should be effective in preventing uplift of the bed

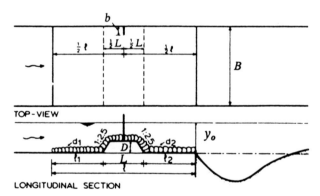

1 Summary of situations

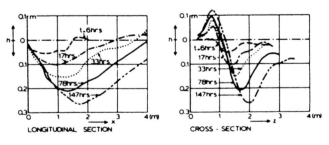

2 Sections through deepest point of scour-hole after different times

Figure 7.9. Definition sketch and some scour hole cross sections.

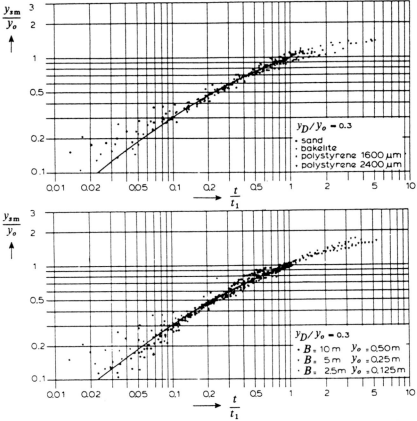

Figure 7.10. Scour development with time for $y_D/y_o = 0.3$ for a range of model scales.

Table. 7.3. Indicative values of α for $L/y_o = 10$ (three dimensional scour case).

	α Bed protection ($L/y_o = 10$) Smooth	α Rough
No sill	1.9	1.6
$y_D/y_o = 0.3$	2.7	2.1
$y_D/y_o = 0.6$	3.7	3.7

protection by flow from the scour hole under reverse flow conditions. The bed protection should exhibit a good filter action in the vertical direction by using sufficient filter layers or artificial filter materials.

All the tests described thus far have been clear-water scour tests. Supply of sediment from upstream reduces scour depths and flattens slopes within the scour hole (Konter

Scour below low head structures 137

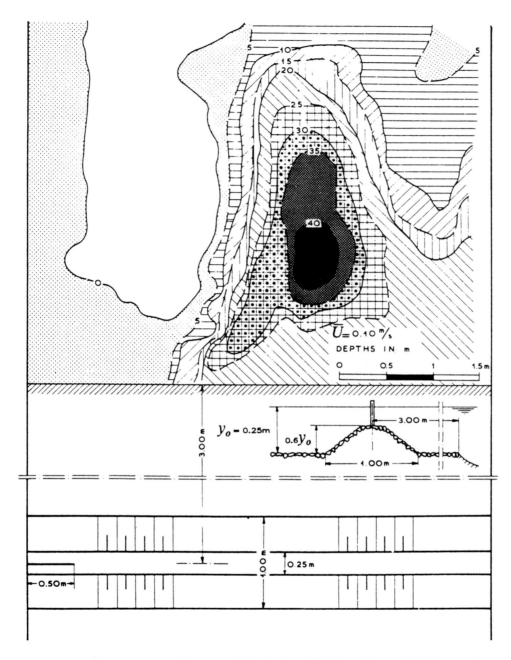

Figure 7.11. Elevation contours in a scour hole ($y_D/y_o = 0.3$, $t/t_1 = 1.0$, rough bed protection).

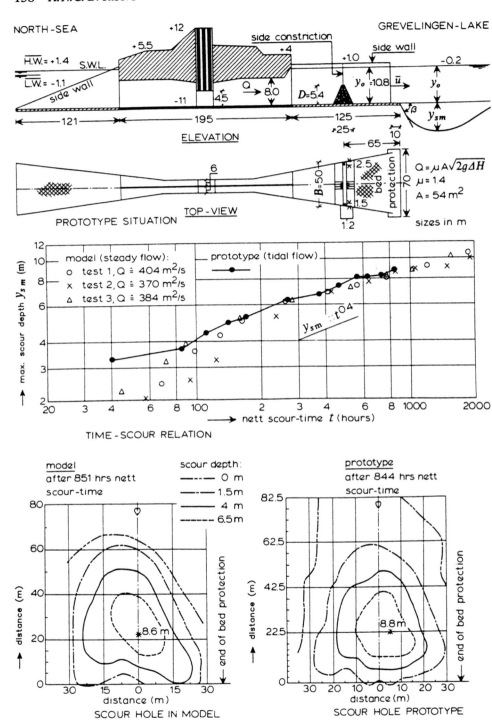

Figure 7.12. Geometric lay-out and measured scour depth. Model-prototype comparison.

and van der Meulen, 1986). If the sediment supply is known, the reduction can be computed by assuming that equilibrium is obtained when the rate of change of the volume of the scour hole in the clear water situation is equal to the upstream supply rate. Scour depths for clear-water give a save estimate because they are the maximum values.

Comparisons of field data with model test results are difficult to make because natural bed sediments often vary substantially, e.g. the occurrence of silt layers in sand. To gain an appreciation of the extent of differences, field scour tests were performed under controlled conditions (de Graauw and Pilarczyk 1981) and repeated in a small scale model (scale 1:30 with polystyrene bed material) using similar ratios of U/U_c. The lay-out and some results are presented in Figure 7.12. These tests show that a very good correspondence between model and prototype data is possible by means of correct modelling techniques.

Farhoudi and Smith (1982, 1984) carried out an extensive study of scour below a spillway (Figure 7.13). The results were in general agreement with Breusers' work, although the exponent in Equation 7.10 differs, and were correlated by:

$$\frac{y_{s\max}}{y_o} = \left(\frac{t}{t_1}\right)^{0.19} \qquad (7.18)$$

where t_1 was found to vary with $\Delta^{1.4}(U_{\max} - U_c)^{-3} y_o^{2.1}$ for constant Froude number. Four sand sizes (0.15, 0.25, 0.52 and 0.85 mm), two bakelite sizes (0.25 and 0.52 mm) and three dam heights (0.1, 0.2 and 0.4 m) were employed. Scour profiles were found to be quite similar at all values of t. The length of the scour hole increased with a lower tail water depth y_o.

Information on scour in *cohesive* soils is scarce. Kuti and Yen (1976) report some tests on scour below a spillway for 6 spillways with varying percentages of fines < 60 µm. The extent of scour decreased with an increasing percentage of fines and a decrease in the voids ratio.

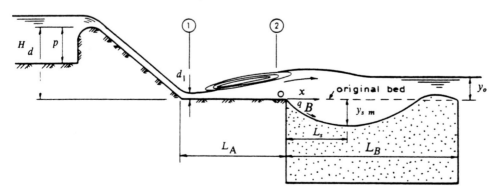

Figure 7.13. Definition sketch of scour below a spillway.

Recommendations for design

The literature on scour in fine sediment below hydraulic structures generally shows that, for a given flow geometry, similarity of scour hole shape is obtained for equivalent values of y_{smax}/y_o. Consequently, results of model tests can be extrapolated to prototype scale provided the time scale of the process is specified. The time development of the scour hole can be described by

$$\frac{y_{smax}}{y_o} = f\left(\frac{t}{t_1}\right) \tag{7.19}$$

where $f(t/t_1) = (t/t_1)^p$ for two-dimensional scour. Values of p range from 0.2 (scour below spillways) to 0.4 (scour downstream of a horizontal bed). More complicated functions were required to describe the three-dimensional case. Time scale can be correlated by:

$$s_t = s_{y_o}^2 s_\Delta^{1.7} (s_{U_{max} - U_c})^{-4.3} \tag{7.20}$$

where s = scale ratio (parameter value in prototype divided by parameter value in model) and $U_{max} = \alpha U$; α depends on the local geometry and ranges from 1.5 (two-dimensional horizontal rough bed) to 4.0 (three-dimensional situations characterized by large horizontal velocity gradients).

As yet no general expressions are available for predicting equilibrium scour depth. An indication is given by the equation of Dietz (1969):

$$\frac{y_{smax(equil)}}{y_o} = \frac{U_{max} - U_c}{U_c} \tag{7.15}$$

However, values predicted by Equation 7.15 seem to be too high, and the time required to attain this depth in the prototype situation is extremely large.

Sediment supply from upstream and/or the presence of more resistant layers will reduce both the maximum depth and the scouring rate, but results to quantify this process are not available. The regime method can be used to give a rough estimate for cases with appreciable supply from upstream such as weirs on rivers. Blench (1957) presented an equation for predicting scour downstream of stilling basins (which contain a hydraulic jump) as:

$$y_s = (0.75 \text{ to } 1.25) y_{2r} \tag{7.21}$$

where y_{2r} is the regime depth in two-dimensional flow (see Chapter 4). Model tests can be used to obtain design information provided geometric similarity is ensured and the bed material is scaled so that

$$\frac{\text{scale } U}{\text{scale } U_c} = \frac{s_U}{s_{U_c}} = 1 \tag{7.22}$$

REFERENCES

Blench, T. 1957. *Regime behaviour of canals and rivers*. London, Butterworth.
Breusers, H.N.C. 1966. Conformity and time scale in two-dimensional local scour. *Proc. Symp. on model and prototype conformity, Hydr. Res. Lab., Poona, India*; 1–8.
Breusers, H.N.C. 1967. Time scale of two-dimensional local scour. *Proc. 12th IAHR Congress, Ft. Collins* 3; 275–282.
Catakli, O. et al. 1973. A study of scour at the end of stilling basin and use of horizontal beams as energy dissipators. *Proc. 11th Int. Congress on large dams, Madrid 1973*, Q41 R2; 23–37.
de Graauw, A.F.F. 1983. *A review of research on scour*. Delft Hydraulics Lab. Rep. S562 (in Dutch, not available).
de Graauw, A.F.F. & K.W. Pilarczyk 1981. Model-prototype conformity of local scour in non-cohesive sediments beneath overflow dams. *Proc. 19th IAHR Congress, New Delhi* 5; 7–16.
Delft Hydraulics Lab. 1972a. *Influence of bed protection on two-dimensional scour*. DHL -report M 847-I (in Dutch, not available).
Delft Hydraulics Lab. 1972b. *Systematic research on two- and three-dimensional scour*. DHL -report M648/863 (in Dutch, not available).
Delft Hydraulics Lab. 1972c. *Influence of bed protection on three-dimensional local scour*. DHL -report M847-III (in Dutch, not available).
Dietz, J.W. 1969. Kolkbildung im feinem oder leichter Sohlmaterialen bei strömenden Abfluss. *Mitt. Theodor Rehbock Flussbaulab.*, Karlsruhe, Heft 155; 1–119.
Eggenberger, W. 1944. *Die Kolkbildung beim reinen Überströmen und bei der Kombination Überströmen-Unterströmen*. Mitt. Versuchsanstalt für Wasserbau. ETH Zürich Nr. 5.
Farhoudi, J. & K.V.H. Smith 1982. Time scale for scour downstream of hydraulic jump. *Proc. ASCE* 108(HY10); 1147–1161.
Farhoudi, J. & K.V.H. Smith 1985. Local scour profiles downstream of a hydraulic jump. *J. Hydr. Res.* 23(4); 343–358.
Ghetti, A. 1954. *The study of bed erosion at weirs by means of small-scale models* (in Italian). Univ. of Padua, Inst. of Hydraulics, Res. Paper no. 167.
Hartung, W. 1957. Die Gesetzmässigkeit der Kolkbildung hinter überströmten Wehren. Diss. T.H. Braunschweig.
Jaeger, Ch. 1939. Über die Ähnlichkeit bei flussbaulichen Modellversuchen. *Wasserwirtschaft und Wassertechnik* 34, No 23/27.
Konter, J.L.M. & T. van der Meulen 1986. Influence of upstream sand transport on local scour. *Proc. IAHR Symposium on scale effects in modelling sediment transport phenomena, Toronto, August 1986*; 208–220.
Kotoulas, D. 1967. Das Kolkproblem unter Berücksichtigung der Faktoren Zeit und Geschiebemischung im Rahmen der Wildbachverbauung. Diss. T.U. Braunschweig.
Kuti, E.O. & Chen-Lien Yen 1976. Scouring of cohesive soils. *J. Hydr. Res.* 14(3); 195–206.
Li, Wen-Hsiung 1955. Criteria for similitude of scour below hydraulic structures. *Proc. 6th IAHR Congress, The Hague*, paper C4.
Müller, R. 1947. *Die Kolkbildung beim reinen Unterströmen und allgemeinere Behandlung des Kolkproblemes*. Mitt. Versuchsanstalt für Wasserbau. ETH Zürich Nr. 5.
Novak, P. 1955. Study of stilling basins with special regard to their end sill. *Proc. 6th IAHR Conference, The Hague*, paper C15.
Novak, P. 1961. Influence of bed load passage on scour and turbulence downstream of a stilling basin. *Proc. 9th IAHR Conf., Dubrovnik*; 66–75.
Raudkivi, A.J. 1963. Study of sediment ripple formation. *Proc. ASCE* 89(HY6); 15–33.
Raudkivi, A.J. 1967. *Loose boundary hydraulics*. Oxford, Pergamon Press. Second edition 1976.

Schoklitsch, A. 1932. Kolkbildung unter Überfallstrahlen. *Die Wasserwirtschaft*, p. 341.
Schoklitsch, A. 1935. *Stauraumverlandung und Kolkabwehr*. Julius Springer, Wien.
Schoklitsch, A. 1949. *Handbuck des Wasserbaues*. 2e Aufl. Springer Verlag, Wien. Bd. I.
Shalash, M.S.E. 1959. Die Kolkbildung beim Ausfluss unter Schützen. Diss. T.H. München.
van der Meulen, T. & J.J. Vinjé 1975. Three-dimensional scour in non-cohesive sediments. *Proc. 16th IAHR-congress, Sao Paulo* 2; 263–270.
Veronese, A. 1937. *Erosion de fond en aval d'une décharge*. Univ. de Padova.

Subject index

abutments 55–59
armouring 14–15, 68, 79–80
bed forms 20–24
bed protection 91–94, 112–113, 134–136
bridge piers 61–96
clear-water scour 3, 65–74, 136
cohesive soils 17, 115, 139
confluence scour 39
constriction scour 3, 42–44, 61
culvert outlets 109–116
design guidelines 46, 54, 58, 88, 107, 116, 140
end sill 125
equilibrium scour 62, 64, 67, 71, 76, 77, 84, 123, 133
fall velocity 10–12
field data 94–96, 138–139
general scour 2, 37–41
horse-shoe vortex 63–64
initiation of motion 13–17
jets
    horizontal 102–103
    plunging 116–121, 124

submerged 99–109
    vertical 103–109
live-bed scour 3, 69, 75–83
low-head structures 123
model testing 3–4, 120, 139
overflow 124
pier scour, effect of
    angle of attack 72, 87
    flow depth 70–71
    flow velocity 75–77
    layered sediments 80–83
    pier alignment 72
    pier shape 70, 73, 74
    pier size 68–70
    pile groups 83–87
    sediment grading 66–68, 78
    sediment size 68–70
plunge pools 118
regime method 25, 52–53
rivers
    bend scour 42
    general scour 37–41
    planform 45–46

roughness
    alluvial channel 17–27
    effect on scour 130–135
sediment
    characteristics 7–10
    gradation 10, 66–68, 78, 115–123
    transport 27–33
Shields' curve 13
slope of scour hole 133
spillways 116–121, 139
spur dikes 51–55
stilling basins 123, 126
three-dimensional effects 127, 134, 135
time scale of scour 129, 132, 140
time-dependent scour 70, 112, 127–140
turbulence, effect of 64, 129–140
underflow 124
wall jets 101
weirs 123